We are the Martians

Giovanni F. Bignami

We are the Martians

Connecting Cosmology with Biology

 Springer

Giovanni F. Bignami
President of the Italian Institute for Astrophysics
Roma
Italy

Authorized translation by Marialuisa Bignami from Italian language edition published by
Zanichelli: "I marziani siamo noi" Copyright © 2010 Zanichelli editore S.p.A

ISBN 978-88-470-2465-6   ISBN 978-88-470-2466-3 (eBook)
DOI 10.1007/978-88-470-2466-3
Springer Milan Heidelberg New York Dordrecht London

Library of Congress Control Number: 2011940856

Springer is part of Springer Science+Business Media (www.springer.com)

For Dindi, for the roots of our culture.

This book has been written for all those people that like to be aware of the things that surround them and who wish to acquire effortlessly an elementary and exact picture of the state of the universe.

Camille Flammarion, *L'astronomia popolare*, 1880

Italian literature is going through a time of general empty bombast. My only fear is to get somehow confused with the ever-present bombastic authors.

Italo Calvino, letter to Giambattista Vicari, 13/5/1964

Preface

Italian secondary school children used to have study notes prepared and published for them by a teacher who had been in the past a school master in my own glorious grammar school; this man's name also happened to be 'Bignami', although no relation of mine. If one of his slight grey volumes could condense the whole of Dante's *Divine Comedy*, then I feel confident I shall be able to concentrate on the present booklet of cosmic history from Big Bang to the rise of life and human beings.

In the present case, we are in need, however, of an interdisciplinary volume, because, first of all, we must build this universe with its bricks of matter and energy, and with stars, galaxies, planets and all the rest. We thus have to speak of physics, astronomy, chemistry and biology.

We shall also have to make the bricks of life and try to understand how it started, with us and elsewhere. Here, a very recent discipline comes into the picture, which we shall call *contact astronomy*; but we shall also need more conventional astronomy, both from the ground and from space, in order to look for other "right" worlds around the "right" stars, where there might be life. We shall then speak of system chemistry, of synthetic biology and (very little) of genetics.

We shall ask ourselves if, by any chance, there may be somebody out there. We shall see how, for a long time, we have imagined alien forms of life and in recent times also to look for them, by listening. We shall thus accomplish a bit of history of science (and of science-fiction as well), but we shall also speak of a still nameless discipline, still to be invented: the way to communicate with somebody we do not know and whom we do not know what to say.

In the end, we shall throw ourselves into some futurology, by imagining what remains to be discovered along the way that binds us to Big Bang.

At the centre of everything, however, there is always the problem of life. Outside Earth we have not found it yet: up to now we have only been imaging it. And about life on Earth too, gaps remain, still missing bits of *fil rouge*, although the latter are growing shorter and shorter.

Working on these student notes, I have discovered there are four ways to try and understand something on the presence—and hence on the origin—of life in the universe:

1. Finding a new far-away Earth, among extra-solar planets, physically unreachable but on which one could "see" unmistakable signs of life.
2. Study the bits of universe that fall on us, meteorites, with all their messages in organic chemistry.
3. Going around exploring, "scratching" the bodies in the Solar system, to see what one can find in it.
4. Trying to understand how life began on Earth, that is to say, the only place where life exists for sure.

The four lines of research are very different from one another as far as objectives and methodology are concerned; they all require quite a lot of mental athletics and open-mindedness. Nobody, I believe, can hope to really master all these topics, as well as the others of which we shall speak, and the present author is surely no exception. If I have tried to push myself beyond my strictly professional competence it is because I believe that at least one attempt at setting the general problem should be performed.

Summing up, with the present notes I would like to contribute to a global vision of the universe, as if it were a forest observed in its entirety. As a physicist and an astronomer, I tend to study single trees (sometimes even twigs, also very far away...), but I am also deeply convinced that science, with its requirements of objectivity and rationality, is the fittest tool to see and understand the forest, vast as it may be.

The main goad for this work has been the contact with the public. In the past two years—the period of incubation for the following pages—I gave above one hundred lectures and seminars in Italy and Europe to a vast and varied audience: secondary school students and teachers; village libraries and Rotary Club lunches; private companies, communicators; and so on.

Also many of the viewers of the series "The Secrets of Space with Bignami", on National Geographic Channel, wrote to me. I answered almost everybody. The remaining people, with my apologies, will certainly find their answers on these pages.

After these 2 years of full immersion in which I gave public talks on science, I feel like asking myself as well as readers: can the statistics be true that I see quoted by Richard Dawkins? According to this famous evolutionary biologist, 44% of US nationals are convinced not only that God created human beings, but that He created them similar to what they are now and that this happened some 10,000 years ago.

Forty-four percent, almost half the population? It is hard to believe. As it is hard to believe that for one Italian out of four (24%, it would appear, the highest percentage in Europe) the Earth takes one month to revolve around the Sun. These dates seem to me impossible to be accepted, or maybe I was lucky enough to have a learned and passionate public, albeit with some unavoidable simple-mindedness.

In any case, if Dawkins were right in being so pessimistic on the general level of scientific knowledge, well, then some interdisciplinary notes on the universe it seems to me might come in quite handy.

Milano, Italy   Giovanni F. Bignami

Contents

Acknowledgments

For a volume of student notes, all-inclusive and interdisciplinary like this one, I required a lot of help, substantially as well as formally. Giulia, to begin with, patiently tutored me in elementary chemistry: I simply tried to pay attention.

Elio Mattia, from IUSS at Pavia, is the mind behind recent views on the origin of life contained in Chap. 7 and in Coda. Ideas are his, I only added possible mistakes (not too many, I hope). I am deeply thankful to Elio for offering me a new cultural dimension, full of rigour and imagination at the same time. My only wish is that Italy will not lose him.

To everybody, students included, at IUSS, which is my university, I owe the stimulus to interdisciplinarity and to Gerda Horneck (of DLR in Cologne) I am grateful for coming over to teach us astrobiology. The part which is most astrophysical in this book owes a lot to colleagues in the Paris P21 Group, "Physique des deux infinis", who have taught me to go from the infinitely large to the infinitely small (and vice versa).

Luca Sivo and his gang at National Geographic Channel Italia contributed quite a bit towards making my ideas and my writing far clearer, also by shaping the TV version of the story. It is an exciting and formative experience, to be suggested to those who have an excessive tendency to sound like a professor.

For style (but also for substance) I was lucky enough to have two not only keen-eyed, but also sensitive and affectionate friends: Lisa Vozza e Federico Tibone, the scientific responsibles for the series. Working with them has really been a privilege not only for their competence and experience, but also for their good disposition. I deeply thank them.

Some friends read the manuscript, with amusing reactions. Among the most alert Sandro Iacchia, a disenchanted and competent physicist: puffing and panting, while we ran, I would tell him what I had in mind, and he would give me precious advice that has ended up in the text.

Cristina Bellon, a science-fiction fan, offered elegant and useful remarks, which also immediately went into the book.

I am also grateful to Alessandro Broi, my webmaster, who knows everything about communication and is forever trying to teach me.

Overture

Man vs. the Universe—The match

Most of us live in bright-lit cities from where we cannot see the starry sky. But on our holidays we may still occasionally find ourselves outside on a very dark night, on a beach with no lights, in the middle of the sea, in a desert or on top of a mountain. Then, if the sky is clear, we will see with our naked eye that it is literally brimming with stars. And with even a small pair of binoculars, we will see the stars miraculously multiplying, whitening a sky that had seemed black.

Lie down on the ground and look up for a few quiet minutes. You will become conscious that the stars really do exist. They are not a figment of the imagination of astronomers and poets. They cannot be dismissed, however distant and unreachable they might be.

Today we know that these stars are more or less like our Sun. They appear smaller only because they are much further away.

You might try to imagine whether, like our Sun, they also have planets orbiting around them, and, if so, what they might be like. Then, inevitably, the question arises: is there life on these planets? At which point you have to ask yourselves: what would scare us most? To know that we are alone in the entire Universe? Or to know that there is someone else out there?

I have asked this question to hundreds, maybe thousands of people, of various walks of life and nationalities. I conduct show-of-hands polls in lectures and meetings. I ask my college students. I ask friends and acquaintances, or even people I meet for the first time. Nearly everyone has an answer. Everyone understands that *tertium non datur* (either we are alone, or there is someone else); very few remain indifferent or refuse to express an opinion. And the overwhelming majority opinion is in favour of there being someone else out there.

The question, it must be stressed, does not specify exactly who—or what—the "someone else" would be. The people I have polled may imagine a vile bacterium, a disgusting spider, a sly monster spreading destruction, or a good fairy dispensing well-being and happiness. It does not matter—anything seems preferable to the angst of loneliness. We simply do not want to be on the only inhabited planet in our Universe.

The bad news, of course, is that we are not yet able to determine whether or not there is someone out there. We have yet to find evidence of life outside Earth, but on the other hand, we also cannot prove that it is not there.

Undeterred, we are still looking, with the best tools that modern astronomy can provide. In the last few years there has been important progress, both in observations and in theoretical studies. The feeling is that we are getting closer to a result. By "result" we mean primarily the discovery of some form of life beyond Earth. Many think that such a discovery would be the most extraordinary event in the history of mankind, so any claim would have to be very convincing. As Carl Sagan said, "extraordinary claims require extraordinary proofs".

This book is written as homage to the majority of people who prefer that we are not alone in the universe. It is written because very recent scientific studies of the Universe, and of life, have delivered results which have changed the very way in which we look at the problem. They are worth recounting.

Another reason for writing this book is to remind readers that, as we look outwards, we should always be aware of our true place in the Universe. This is important because today's culture, with its widespread ignorance of science, or rather of the history of science in the last four (or forty?) centuries, tends to revive anthropocentric irrationality.

As a light-hearted introduction to the discussion that follows about the origins of the Universe and of life on Earth, and perhaps outside Earth—that is the origin of man in the Universe—I am inserting here the broadcast of the second round of the "Anthropa Cup", a football challenge of two games between the teams Dinamo Universal F.C.

and Man-Centred United, which I received in a zipped file from an anonymous source.

The anonymous source reminds us that the first round, played many centuries ago in the "Mediterranean" stadium, ended with a resounding 2–0 victory for Man-Centred United. The goals were scored by Aristotle and Ptolemy, the philosophers who place Man and Earth at the centre of the Solar System, therefore of the Universe. They were the forefathers of anthropocentrism, which allocates to man and his planet a special place in the universe.

The second round, naturally, is played in the "Universal" stadium. Here is the broadcast.

A huge crowd. Billions of stars and a billion of galaxies are silently cheering for Dinamo Universal, and also provide the lighting for the night game.

The rowdy Man-Centred supporters crowd the South Terrace with their collective trillions of brain cells (incredible as it may sound, each fan is endowed with a hundred billion), and there are also a bunch of hooligans dressed in the cardinal-red Man-Centred football shirt.

The starting whistle is around 500 AD. The Man-Centred game plan is immediately revealed: Aristotle and Plato, at centre field, emboldened by their initial strong advantage, are wasting time. They kick the ball round and have the Sun and all the planets turn around the Earth. But all of a sudden the Dinamo-Uni's Polish stopper, a certain Kopernik, comes into action. Until now he has been an unknown semi-illegal immigrant in Italy, but a new graduate of Ferrara University. Kopernik takes possession of the ball. He kicks a deep ball which slices through the opponent's defence. It is a revolutionary book (not for nothing called De Revolutionibus Orbium Coelestium), which in 1543 for the first time puts the Sun at the centre with the Earth rotating around it.

Kopernik's pass is picked up by the legendary Galileo Galilei who plays with a small telescope stitched on his Dinamo United midnight-blue football shirt. Galileo substituted Giordano Bruno, sent off field by the referee in a burning lack of justice (the referee has a hard time running, as he has to pick up the heavy cassock that hides his shorts. . .).

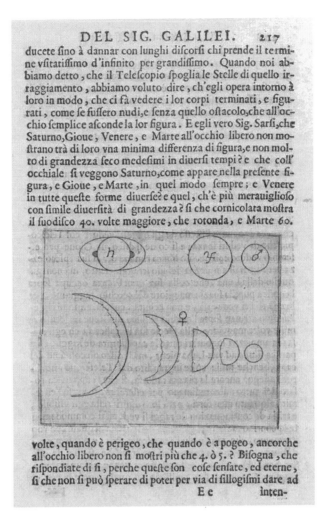

DEL SIG. GALILEI.   217

ducete fino à dannar con lunghi difcorfi chi prende il termi-
ne vfitatiffimo d'infinito per grandiffimo. Quando noi abbiamo detto, che il Telefcopio fpoglia le Stelle di quello ir-
raggiamento, abbiamo voluto dire, ch'egli opera intorno à
loro in modo, che ci fà vedere i lor corpi terminati, e figu-
rati, come fe fuffero nudi,e fenza quello oftacolo,che all'occhio femplice afconde la lor figura. E egli vero Sig. Sarfi,che
Saturno,Gioue, Venere, e Marte all'occhio libero non mo-
ftrano trà di loro vna minima differenza di figura,e non mol-
to di grandezza feco medefimi in diuerfi tempi? e che coll'
occhiale fi veggono Saturno,come appare nella prefente figura, e Gioue, e Marte, in quel modo fempre; e Venere
in tutte quefte forme diuerfe? e quel, ch'è più merauigliofo
con fimile diuerfità di grandezza? fi che cornicolata moftra
il fuodifco 40. volte maggiore, che rotonda, e Marte 60.

volte, quando è perigeo, che quando è a pogeo, ancorche
all'occhio libero non fi moftri più che 4. ò 5. ? Bifogna, che
rifpondiate di fi, perche quefte fon cofe fenfate, ed eterne,
fi che non fi può fperare di poter per via di fillogifmi dare ad
                              E e   inten-

Fig. 1 At the bottom of the page from *Il saggiatore* (1623) the phases of Venus drawn by Galileo can be seen. This observation demonstrates that Venus orbits around the Sun; the planet appears to be "full" when it lies on the opposite side of the Sun in relation to the Earth. This image comes courtesy of the Photographic Archive of the Museo Galileo in Florence

Galileo pushes to the centre, holding on to his telescope and kicks into the net Kopernik's brilliant pass by observing satellites around Jupiter and the phases of Venus. From now on the Earth is not at the centre of the solar system, and the planets are revolving around the

Sun. It is no longer a mere abstract mathematical theory which the Church had not even bothered to blacklist (it will do so only in 1616, after Galileo's score): now it is a safe and undisputable astronomical observation.

In spite of the frantic waving of the banner "Bellarmino, you take care of him" by the hooligans on the South Terrace, Galileo scores the first, awesome goal that marks Dinamo United's comeback.

In the ensuing scuffle in the goal zone, however, Galileo himself is cautioned by the referee. He is forced to chew and swallow the ball and even to say that he enjoyed it.

The game re-starts, but Galileo will never be the same. In the "After-Match Trial", which took place—admittedly with some delay—in 1984, Pope John Paul II had to admit that Galileo should not have been punished, and had not actually even committed a foul.

At the end of the first half—we are now in 1859—Dinamo Uni brings in a powerful English centre-forward, Charles Darwin, who, despite his advancing years, asserts himself on the Man-Centred half field. He is dressed as a monkey, but then throws his mask and displays the fish-creature from which we all descend. He scores!

Not only is man no longer at the physical centre of the universe, but also now we know that we only are a kind of evolved fish. The referee would like to disallow the goal, but cannot find in the regulations a ban specific about scoring while holding a fish.

Two-nil for Dinamo at the end of the first half. The Man-Centred team goes to the locker room with their heads down, followed by their cardinal red-clad masseurs offering incense as a pick-me-up.

The deciding second half starts around the mid-1900s. Dinamo United can now pick fresh talents from the newly available fields of nuclear and theoretical physics. One such talent is the classy English player Sir Fred Hoyle, even if bespectacled and a bit overweight. With his "theory of nucleosynthesis" he is building up a new offensive from centre field. Hoyle knows how the stars work and he understands why they burn. He barges into enemy territory wearing Mendeleev's table on his midnight-blue shirt (the referee is clueless as to what this table is, so he lets it be. . .).Than he points at the Table so as to let everyone know that he has discovered that the chemical elements are made by stars. From their own terraces the stars scream to the men on the south terrace: "Dust, you are all but our dust!" Hoyle is carried away by the excitement and he scores!

An important goal for Dinamo United. The matter from which man is made (and the Earth and everything else) comes from the Universe, and has really nothing special to it. The stars cheer, hugging themselves into spontaneous spirals. Three-nil. Things are looking bad for Man-Centred.

Last frantic minutes of the game, from 1990 to today. Astrophysics is now the breeding ground of fresh talents, so many that it is impossible to name them all. Dinamo Universal takes a two-pronged offensive: the composition of the universe matter and the planets surrounding other stars.

The new hires from the astronomy pool find out quickly that the matter that makes us (and all we see around us) is nothing but a pinch of salt in the soup of the universal matter. Most of the universe, precisely 96%, is made of matter or energy that we call "dark", just to give it a name, but have nothing to do with ordinary matter. In other words, we belong to a trivial 4% of the universe.

Here is another great score against Man-Centred: not even our own matter is central in the universe. Four-nil.

The next offensive comes from a combined action of Dinamo's two wing players: Earth Astronomer and Space Astronomer.

Many people had always believed that planets exist around other stars—from Giordano Bruno, the one of the scorching send off, to Giacomo Leopardi. But no one had ever seen one until 1995, when the first extra-solar system planet was spotted. Today we have counted nearly 1000 and within a few years, the count will be in the thousands. Planets, it turns out, are the norm, not the exception, around stars. Our own Solar System is just one of many, nothing special about it. Fifth goal for Dinamo Universal, and the referee blows the final whistle.

A resounding five-nil win, which crushes the 2-0 of the first round and decides the winner of the Anthropa Cup, which is delivered with great pomp to the victorious Dinamo.

Coach Universinho immediately brings in a huge steamroller which, in a theatrical gesture, flattens and destroys the cup. Farewell anthropocentrism.

The stars, planets and galaxies scream like crazy (some feel an obscure, indescribable, unknown presence in the air. . .). Some disrespectful voices scream: "Make Giordano Bruno a saint now". They are soon quieted.

The referee responsible for Bruno's send-off and Galileo's heavy caution, not to mention the clumsy attempt to deny Darwin's goal, runs for the locker room and escapes through an underground tunnel dug under the river Tiber.

This is the commentary so far (thanks to the anonymous sender of the zipped file).

The rest is history, as they say. I will try to tell the story in a less frantic manner in the next chapters. I will cover the origin of the universe, of the stars and the galaxies, of the elements of which we are made and all the rest. I will also cover the birth of the planets and the molecules that float in space, by themselves or carried by interplanetary and interstellar rocks and ice balls.

Then I will talk about the new astronomy of the third millennium, that is, "contact astronomy" which started with meteorite collecting and today allows us to visit planets, comets and asteroids to have a taste of them in situ. There we find important molecules, "the bricks of life", amazingly similar to ours, arousing the suspicion that like atoms and molecules, we may also come from outside the Earth. So are we the real Martians?

Naturally the distance between "bricks" and "life" is enormous, and it is not resolved. A truck full of bricks is not enough to make a house. Life is a crucial step, still unknown, that so far defies everyone's understanding, a bit like the Big Bang genesis (although we now know quite well what happened after the Big Bang).

There is no scientific evidence that life comes from outside. However, I will also talk about "panspermia", that is the possibility that elementary forms of life survive in space by travelling between planets (someone has even suggested between stars), carried by Enterprise-class meteorites. No, we are not going to talk about the famous Star Trek Enterprise, nor the spaceship carrying the cute E.T. Let us use our imagination, but let us also try to stick to the facts. And on this note, if there really is someone out there, might they be trying to contact us? I will recount the half-century history of the SETI project, the search for extra-terrestrial intelligence. We have not found it yet; but, as I will explain, we ourselves are sending confusing messages into space, a sort of message-in-a-bottle inside what we will call the "Berlusconi Bubble" that now has engulfed thousands of stars.

Chapter 1
Let's Make the Universe: Light and Matter

We don't know what was there before Big Bang. I'll say it right away, because it is the answer to one of the questions I get asked most often. Or rather, let's say that maybe it makes no sense asking ourselves what was there. Or let's say it is of no interest to us, but only because we do not know what answer to give. In the end, let's forget it.

Those who believe, naturally see in it the work of a Creator, which at this initial stage still cuts across all religions. Some free-thinking theologians label this solution "Stopgap God", called upon somewhat on purpose. They may not be wrong. Those that are not so lucky as to believe, maybe because they think that a Creator is no scientific notion, on the contrary simply don't know and know they don't know.

A scientist, at the most, upends the problem: if you ask a believer what existed before God, he will answer you correctly: "But God is eternal, He was always there". And then, you will say, the universe too could always have been there. In fact some cosmologies have imagined pre-Big Bang universes which, somehow, subsequently generated Big Bang and thus created our particular universe. Another solution? Who knows? Certain it is that the presence of other universes, unknown and inaccessible to us, could also be useful in other contexts, as we shall see further on.

On the contrary, what counts very, very much for our story is what happened starting with Big Bang and afterwards, up to today. It is useful to understand that we, human beings of today, are the children of Big Bang. Not in a vague or abstract way but in a very matter-of-fact way, in the Big Bang of some 13.7 billion years ago, what happened was not something outside us, something that did not

G.F. Bignami, *We are the Martians*,
DOI 10.1007/978-88-470-2466-3_1, © Springer-Verlag Italia 2012

concern us. On the contrary, through Big Bang, the matter was created, of which all of us are made; everything we can see around the universe by means of our eyes or by instruments was created too. Nay, of Big Bang we carry within us tangible marks, almost a brand mark.

The First Instants

What may have happened just at the beginning of the universe, nobody (yet) knows. But right afterwards, very little time afterwards, we do know; today we can at least describe what was happening. Nay, we have brought back a very brilliant idea, more than a century old, to state how big was the new-born universe: its diameter corresponded to the so-called *Planck length*, which, if expressed in metres, is worth some 10^{-35} m; it is a measurement of length so incredibly small that one hundred thousand billions of billions of billions would be necessary to make up a millimetre.

This length is a shrewd combination of the fundamental constants of nature: light speed in Einstein's theory of relativity, the constant of universal gravitation in Newton's theory and Planck's constant from quantum mechanics.

By employing these constants, Max Planck built a system of unities of measurement (of length, time, mass, electric charge and temperature) that he presented to the Academy of Prussia (Prussian Academy) in May 1899.

Planck knew well that the unities of measurement that we use every day (metre, second, kilogram and so on) are somehow connected either with human beings or with the planet Earth. The unities he suggested, on the contrary, were totally independent of human beings; they were rather connected with fundamental and thus universal laws of nature. As Planck himself told his Prussian colleagues: [Such constants] maintain their meaning for all times and all cultures, even extraterrestrial and non-human, and can, therefore, be called "naturel units".

Thus, he presented the natural unity of length, that is, the above-mentioned Planck length. To cover it, light takes up a very short time, some 10^{-43} s, which is *Planck time*.

The good thing of the Planck system is that inside its unities, all the physics is contained that we can imagine, all pressed together, exactly in that "unified" way which today we can no longer repeat.... To describe what happened from time "zero" to Planck time, that is to say, in the first 10^{-43} s of the life of the universe, does require a theory which would be able to describe, really at the same time, all the forces in the universe. In order to develop this theory, we are employing all sorts of things, the great particle accelerators included, such as CERN, but as yet to no avail. Therefore, we shall not try to tell how the physics of the first 10 s was, although we know there must have been a physics and that it must have been beautiful.

What is possibly, up to the present, the best description of the new-born universe, which could all fit into a Planck length, we owe to the incomparable pen of Italo Calvino, in Cosmicomics "All at One Point":

> Naturally, we were all there – *old Qfwfq said* – where else could we have been? Nobody knew then that there could be space. Or time either: what use did we have for time, packed in there like sardines? I say 'packed like sardines' using a literary image: in reality there wasn't even space to pack us into. Every point of each of us coincided with every point of each of the others in a single point, which was where we all were. In fact, we didn't even bother one another, except for personality differences, because when space doesn't exist, having somebody unpleasant like Mr Pbert Pberd underfoot all the time is the most irritating thing. Haw many of us were there? Oh, I was never able to figure that out, not even approximately. To make a count, we would have had to move apart, at least a little, and instead we all occupied that same point. Contrary to what you might think, it wasn't the sort of situation that encourages sociability; I know, for example, that in other periods neighbours called on one another; but there, because of the fact that we were all neighbours, nobody even said good morning or good evening to anybody else.
>
> Italo Calvino – The complete Cosmicomics – Translated by Martin McLaughlin, Tim Parks and William Weaver – Penguin Books Ltd

Matter and Antimatter

After a Planck time, in which temperature was rather high (all of 10^{+32} degrees), the universe, while expanding, began to grow cooler and to create physics the way we know it today: beginning with separating the force of gravity from all other forces.

In the first one hundred billions of Planck times, that is to say, up to the venerable age of 10^{-32} s, the space which contains the universe expands, swelling even faster than the speed of light (mind you: what is expanding is space, not matter...). Thus, cosmos reaches the respectable dimension of a grapefruit. Yes, there was a time in which the whole universe would fit inside the volume of a grapefruit: it seems little, but some 10 cm is a long way if compared to a Planck length.

Later on, much later on, that is to say, during the universe's first millionth of a second, a lot of things happen, among which is the advent of light and of the fundamental bricks that will go to make up today's matter, all this inside a volume just slightly bigger than the Colosseum. And here, the first surprising, almost miraculous, event takes place, one to which we owe our existence, although we cannot understand why it took place.

In the very thick "primeval soup"—Calvino would say a spoon could stand in it—matter was forcibly created in two different forms: the one we are made of and its symmetrical one which we—with a bit of racism—call antimatter.

Antimatter is almost identical to matter, but it has an opposite electric charge as well as an interesting property: when it meets matter, together with it "annihilates" itself, that is to say, it disappears and goes back to being simply energy, in the guise of photons. Big trouble: if from the primeval soup perfectly symmetrical and equal quantities of matter and antimatter had surfaced, the universe would today be a big lake of photons, in which we could not even swim, because we—vile matter—would not exist at all.

For reasons which are not yet clear, however, the primeval soup generated more matter than antimatter: a small excess of about one part to a billion. It appears to be little, but it is exactly this matter that has not "annihilated" itself (because antimatter was already exhausted) and thus has been able to form the universe and then ourselves. Halleluja!

Let us reverently keep in mind this asymmetry, which is not yet explained and to which we owe our life. Also because there are people who believe that to this fundamental asymmetry in our universe another one is bound, that we shall meet further on: the one contained in the "chirality" of big organic molecules that go to make up living matter, another spectacular tie between us and the stars.

Fusion and Nucleosynthesis

Let us go back to the birth of the world. At this point, the first stormy second in the life of the universe has gone by, and, with a bit of luck, light and matter have appeared: that is to say, the famous *protons, neutrons* and *electrons* of which we study in school. The temperature has gone down to a mere billion degree centigrade, and the universe is all contained in a sphere with a radius of one hundred thousand kilometres, corresponding to one-fourth of the Earth–Moon distance.

After the proton, which is the simplest atomic nucleus (the one of the *hydrogen* atom), *deuterium* begins to form, another sort (or *isotope*) of hydrogen which is made up of one proton and one neutron, held tightly together by those nuclear forces that will end up by composing the nuclei of all other elements. Deuterium is still in abundance today—a nucleus of it is found for every 7,000 nuclei of hydrogen—and it is certainly the one which was formed after Big Bang: we have not yet found another way of producing it in noteworthy quantities.

Then the time comes of *thermonuclear fusion*, the one we've been dreaming about repeating for a long time, in a lesser dimension, on Earth in order to produce energy, but until now to no avail. As we shall presently see in our cosmological *fast forward*, as soon as we've been able to make stars, fusion is the very condition that keeps them burning.

The first element different from hydrogen, the very light helium, is born of the fusion of protons: thanks to a high temperature that makes the latter move and clash forcefully against each other, thus overcoming the repulsion due to their electric charge. The nucleus of helium (which physicists call *alpha particle*) is made up of two protons and two neutrons and is very solidly bound together, more so than all its light neighbours in the Mendeleev table. This special stability of its own is fundamental on all that comes next, because it will be reflected in the nucleus which is at the foundation of our existence, *carbon*: for the time being, it is not yet in existence, but when it will be born, it will be very stable exactly because, essentially, it is made up by three alpha particles stuck together.

In the meantime, while the universe keeps growing colder, a bit of *lithium* finds occasion to be generated, with three protons in its nucleus, and maybe a pinch of *beryl* too, with four protons. Thus, the universe, although very young, manages to complete the prodigy chased after by medieval alchemists as well as by modern physicists, generating new matter starting from existing "bricks".

Cosmological, or primeval, nucleosynthesis, which we just described, takes place in the first most famous 3 min of the universe's life. At this point, cosmos has a diameter of almost one hundred million kilometres, smaller than today's Earth–Sun distance; its temperature goes down to almost one million degree centigrade, too low to further support fusion.

Thus ends nucleosynthesis, which had been a real race against time; its outcome, however, still lives today around us, for instance, in hydrogen in ocean water and even inside ourselves. For, in our bodies, there are many kilos of hydrogen atoms (and maybe some original atoms of helium and lithium): their nuclei were formed 13.7 billion years ago; they are our tag reading "made in Big Bang".

The First Atoms and the Pictures of the New-Born Universe

Our understanding of cosmological nucleosynthesis—one of the triumphs of science in the second half of the twentieth century—also produced another important result: we know exactly how much matter, of the sort that makes up our body, exists in the universe at the conclusion of nucleosynthesis. After the three initial minutes stopped or finished, matter of this sort will not be made again. At the most, as we shall see, stars will transform it, but no more of it will be generated.

In the first 3 min, we saw a lot of things happen. In the following 380,000 years, more or less, on the contrary, nothing happened. The soup of photons, protons and electrons, with a little deuterium, helium and lithium nuclei, keeps expanding. But it expands with a special property: despite the presence of so many photons, the *quanta* of light, nothing is to be seen; it is pitch dark.

For, photons have a special affinity with electrons: if the latter go roam free, photons clash against them right away and cannot proceed; that is to say, they don't make light. It is a matter of density and of temperature. During 380,000 years, a large number of free electrons are around, in a relatively small space. And, on top of that, the universe is still so hot that electrons move frantically; thus, although negatively charged, they can't perceive the presence around of a lot of protons, positively charged, that expect nothing less than catching them and binding them to themselves.

On the contrary, this begins to succeed when the universe becomes 380,000 years old, it reaches the size, more or less, that our Galaxy has today (a diameter of one hundred thousand light years) and its temperature goes down to 3,000°C, about a half of the one on the surface of the Sun.

Another admirable event is that electrons, as they slow down, "feel" the protons' charge, get captured and begin revolving around protons, and thus the first *neutral atoms* are generated: mainly hydrogen, with interesting traces of other chemical cosmological elements. At the same time, however, photons are let loose, and the universe can finally be seen.

One of the best results of modern astronomy is the image in Fig. 1.1: a "photograph" of the new-born universe, which depicts it as soon as it could be seen. "Seeing" in this case means looking at a map of the sky that describes how the first photons issued from the universe were distributed.

For this very discovery, first, and then for the detailed representation of those photons, two Nobel Prizes have been given (to Penzias and Wilson in 1978 and to Mather and Smoot in 2006). As we are about to see, however, it took some time before it was understood what type of radiation it could be, that is to say, what should be sought for.

Fossil Radiation

The expansion of the universe, which was discovered when it was observed that other galaxies are moving away from ours, implies a constant cooling down of cosmos: the initial energy disperses in a constantly growing volume. So the photons generated by Big Bang, which began "very hot" 13.7 billions of years ago, are today much "colder". That is, they have much less energy. Their temperature can be measured, and it is found only 2.7 above absolute zero (i.e. at about

Fig. 1.1 This is the "first photograph" of the universe: it shows what the sky looked like only 380,000 years after Big Bang, at the time when cosmos had become for the first time transparent to electromagnetic waves. The *shades of grey* represent tiny fluctuations of temperature (+200 K) as measured by the Wilkinson Microwave Anisotropy Probe in the fossil radiation dating 13.7 billion years ago. Mage Source: NASA/WMAP Science Team

−270°C). This is exactly the value envisaged by the theory of the expansion of the universe, which from measuring this background radiation of the sky has been spectacularly confirmed.

From the beginning, what was later observed (from the 1960s onwards) was only the theoretical anticipation. Ralph Alpher and Robert Herman, two astrophysicists later on disregarded by the Nobel committee, in 1948 had already anticipated that, according to the Big Bang theory, in the "background of the sky", there should have been a low temperature radiation a fossil remnant of the first photons escaped just after Big Bang. They had also anticipated the temperature of this fossil radiation: photons were to be at 5 K, only 5° above absolute zero. The anticipation met with general incredulity.

On the contrary, Alpher and Herman were right, apart from a mistake by a two factor about temperature. Their anticipation has proved an incredible result, for its daring and depth, considering the competence of their time. It is a bit as if we were to anticipate today for the coming 4 April the landing of a flying saucer of 100 m. diameter on Piazza S. Pietro, obviously generating a lot of scepticism. Then, on the set date, however, everybody can observe a flying saucer that actually lands in front of the obelisk, except its diameter is only 54 m.

Let us go back to the image in Fig. 1.1 and let us try to take a close look at it. We can easily see that fossil radiation is not uniform on the whole sky: there are parts where it is more intense (or warmer) than in others, although the difference is only one part on one hundred thousand.

These fluctuations in radiation are interesting: their only possible explanation is that they represent the faithful imprint, the mould of similar fluctuations in the density of matter, of that matter through which photons in the end managed to pass in order to escape, but most of all, of *dark matter*.

By means of this phrase, for lack of better words, we define that "thing" that makes its presence felt by means of gravitational attraction (in particular, by determining the shapes of galaxies and their positioning themselves in masses), but that, unlike stars or interstellar gas, does not emit electromagnetic waves.

Dark matter, which, at the time of Fig. 1.1, would presumably make up the most part of matter in the universe, was not smooth and uniform, but (although so young) would already show small "wrinkles", exactly one part on one hundred thousand. How small these wrinkles are is well explained by an analogy with the Earth: if the surface of the Earth were smooth in the same way, the difference between the highest summit and the deepest ocean depression would be laughable, equal to only 70 m.

And we are lucky there were fluctuations: it was exactly those small variations in density, captured in the photograph of fossil photons, that stopped the universe from being simply a thin soup, tasteless and uniform, of hydrogen and little else. And in this case, once again, today we would not exist: the universe would be all the same, all over the place, and the sensual Mrs. Ph(i)Nko of "All in a point" would not be able to make *tagliatelle* for us. In Calvino's words:

> We got along so well all togheter, so well that something extraordinary was bound to happen. It was enough for her to say, at a certain moment:"Oh, if I only had some room, how I'd like to make some tagliatelle for you boys!" And in that moment we all thought of the space that her round arm would occupy, moving backwards and forward with the rolling pin over the dough, her bosom leaning over the great mound of flour and eggs which cluttered the wide board while her arm kneaded and kneaded, white and shiny with oil up to the elbows; we thought of the space that the flour would occupy, and the wheat for the flour, and the fields to raise the wheat, and the mountains from which the water would flow to irrigate the fields, and the grazing lands for the herds of calves that would give their meat for the sauce; of the space it would take for the Sun to arrive with its rays, to ripen

the wheat; of the space for the Sun to condense from the clouds of stellar gases and burn; of the quantities of stars and galaxies and galactic masses in flight through space which would be needed to hold suspended every galaxy, every nebula, every sun, every planet, and at the same time we thought of it, this space was inevitably being formed, at the same time Mrs Ph(i)Nko was uttering those words: '...ah, what tagliatelle, boys!'
Italo Calvino – The complete Cosmicomics – Translated by Martin McLaughlin, Tim Parks and William Weaver – Penguin Books Ltd

But luckily, fluctuations there were and they were the right size (as it happens). Let's then see how, from them, all the things described by Calvino's pen were issued, from stars to planets to Mrs. Ph(i)Nko's breast included.

The First Stars

Some time has gone by since matter, finally separated and independent from radiation and only made up for the time being of very light primeval elements, began somehow to "condense", that is to say, to follow the evolution of density fluctuations, left to fend for themselves under gravity forces. To be exact, about two hundred million years go by, during which the sky has been growing uniformly cooler. To our eyes, everything would seem dark, because stars had not been born yet. . .

In the dark too, however, gravity was at work, let loose. It was at work on the matter formed up to that point, that is to say, the stuff generated in the first 3 min, that was however dispersed in the much richer sea of dark matter, a sea that had exactly the density variations we have been talking about. Exactly from these wavelets, many billions of years before Isaac Newton came to understand it, could the phenomenon of *gravitational collapse* get on its way: attraction towards the centre due to gravity generates, as in a landslide, spherical concentrations of thicker and thicker matter.

"Ordinary" matter too, 4% of hydrogen and little else, unavoidably slides into the big "gravitational hollows" created by the fluctuations of dark matter, which lords it over everything. As density increases, and thus temperature too, matter splits into condensation centres. At this point, one, one hundred, one thousand, one hundred million tiny inverted Big Bangs take place: falling on itself, taken into its own attraction, matter by contracting forms atom balls, thicker and thicker as well as hotter and hotter.

As density and temperature grow, atomic nuclei, mostly of hydrogen, are pushed with ever increasing force one against the other until, suddenly, a second era of nucleosynthesis begins (which is still going on today). That is to say, thermonuclear fusion starts all over again, this time on a much smaller scale than in the first 3 min; nuclei of helium and other light elements start taking shape, with a lot of energy being liberated. In other words, the first generation of stars has been lit up.

They all have a great mass compared to our Sun, maybe a hundred times bigger. And their initial matter cannot but be mainly hydrogen, with 20% of helium and a pinch of lithium. It's the ideal recipe for thermonuclear fusion, which advances in great sway, almost up to a paroxysm in the stars with bigger mass. For, these stars are short lived, a few tens of millions years, quickly burning up all their "light" fuel. And what is the result of nuclear combustion? Nuclei of ever-heavier chemical elements, constructed according to the laws of nuclear physics.

Thus are generated the heavier nuclei of lithium generated, also the carbon of our tissues, the oxygen we breathe, the calcium in our bones, all the way to iron in our blood. Stellar nucleosynthesis is similar to the cosmological one of the first 3 min, but it is more complete simply because it has much more time at its disposal.

But it is exactly when one reaches iron that Mother Nature demands a stop in this process. For, iron's is a special nucleus. If the nuclei of the lighter elements are synthetized, energy is released, exactly the energy that keeps the star warm (and that gives us a tan). But beyond iron, the energetic balance of the process changes sign: in order to make the heavier nuclei of iron, which we do see in abundance around us, it is necessary to *supply* a lot of energy.

Supernovae: All the Elements Are Born

Mother Nature took care of it, of course. When a star exhausts its fuel, in the sense that it has operated all the nucleosynthesis it could, it does not release enough energy to oppose its own gravitational collapse, as we shall better see further on. Suddenly, in a split second, the star implodes, the matter of which it is made falls on itself centrewards. Immediately afterwards the star explodes: that same matter, heated and compressed in implosion, bounces outwards.

The mechanism of this gigantic explosion, called *supernova*, is intricate in its details, but some of its characteristics can be understood in an intuitive way; they are fundamental for our story, since we are creating our world as it is today and building the base of our existence itself: let us keep this in mind!

First

The energy involved in the explosion of a supernova, which lasts let's say 1 s, can be in the range of 10^{44} J, corresponding to all the energy that our Sun can emit in its whole life. It is an energy by far sufficient to form elements and to complete the Mendeleev table beyond iron. With different quantities and proprieties, issued of the laws of nuclear physics, germanium for our transistors, silver for our forks and spoons, gold for our earrings as well as uranium for our armaments are thus generated. Exactly so, every single nucleus of these elements, which are today found on Earth and that we see every day, was created that is synthetized from (out of) lighter elements in supernova explosions. This is "explosive nucleosynthesis", the third type of nucleosynthesis, after the primeval and the stellar ones, which completes the periodic table and generates the world essentially as we see it today.

Second

The explosion, after generating the nuclei of new elements, disperses them together with the old ones here and there in space. This way an *interstellar gas* is created, thus called because it literally lies among stars. The gas becomes richer and richer in heavy elements with every explosion; the heavy elements are much more useful for the purpose of originating life than the tasteless primeval soup (composed of hydrogen and little else).

Third

From the explosions of supernova, other interesting products derive as well. They are the central hard cores of stars, the over-compressed remains of matter collapsed on itself and not swept outside.

Density, during implosion, plays funny tricks. It can compress great quantities of matter to a point of density similar to the one of the atomic nuclei, turning them entirely into neutrons. Thus, *neutrons stars* are generated, which have a mass on the order of our Sun, compressed into a sphere with a radius not above 10 km. But implosion, if the initial mass is big enough, can go even beyond that and produce the famous *black hole*: an object in which a mass equalling many times the Sun is compressed inside a radius of a few kilometres; thus, it is even denser than a neutron star, nay so dense that the *speed of escape* from its surface, which should be acquired in order to go away from it indefinitely, becomes higher than the speed of light. Not even photons can escape from a black hole, which, exactly for this reason, is black....

Of course, not all the stars in the universe end up so spectacularly. In actual fact, the length of the life and the sort of death of a star depend very much on its initial mass. Those that are born very big, tens or hundreds of times the mass of the Sun (such as "first-generation stars"), burn up everything very quickly; they live a few hundreds or even tens of millions years and end up in a supernova, maybe leaving a black hole in the centre.

Less massive stars, up to some 20 solar masses, can leave behind a neutron star, whereas the ones similar to the Sun, which are much more frequent, do not produce great explosions and have a long life (for instance, ten billion years), and during their whole life, they emit gas enrich by their nuclear activity.

Summing up, what by explosion, what by normal emission, both giant stars and more ordinary ones, such as the Sun, generation after generation, have been enriching the interstellar gas with heavy elements.

Galaxies

We have been seeing that first-generation stars were born because of the fluctuation in density of dark matter in the initial cosmos. Born together, the ones close to the others, stars are bound by gravitational attraction into great conglomerates that can easily contain billions of them, or rather hundreds of billions. Thus, *galaxies* are born that up to a century ago would be called "island universes": so big they would appear each as a self-standing universe.

The hundred billion stars—more or less—contained in each galaxy organize themselves according to gravitational attraction in elegant forms and structures, such as the giant spirals, and revolve majestically slow around the their centre (which is often inhabited by a black hole of a mass corresponding to billions of Suns).

Our own Milky Way, for instance, which is a disgustingly ordinary and middling galaxy, completes a whole revolution around itself every two thousand million years. Since it was born, for sure none much later than Big Bang, it must have taken at least 50 revolutions while its stars would be born, die and rise again.

Speaking of stars and galaxies, let us go to some computing. If a commonplace galaxy (and therefore typical) like ours contains one hundred billions of them and in the universe we know billions of galaxies, the total number of stars in cosmos is really high: 10^{22}. It's really a big number: stars are more numerous than grains of sand on all the beaches and in all the deserts on our planet.

Too bad nobody up to now has ever observed first-generation stars, the ones that a few hundreds of billions years after Big Bang began to generate heavy elements. They are too ancient and thus too far away for our telescopes, but we're working on it.

Chapter 2
Let's Make the Rest of the Universe

In order to win the Olympic Games, one must train, of course, and believe in it, but today, we know that by far the most important requirement is choosing your parents well: you need parents that have the right genetic code, naturally bent to athletic excellence. In the same way, if a star aims at being eventually surrounded by a good planetary system, maybe made up of planets ready for life, what really counts is that it should have chosen well the cosmic gas-and-dust cloud out of which to be born.

Which are the "right" clouds? It's certainly not the first-generation ones, direct heirs to the condensation of primeval gas masses. Rather, it would be better for other stars to have had time to enrich that gas with the precious elements necessary to life. In other words, it would be better to choose a middle-aged cloudlet, in which many generations of stars may already have laid everything which we need– and it is a lot more then primeval gas.

What's in the Sky Between Star and Star?

If we examine in detail what our body is made of, we shall find that it contains a collection of samples from the Mendeleev table of chemical elements, albeit in highly variable quantities from one element to the next. For sure, to begin with, a lot of CHNOPS is needed. It's a bar-room name for the cocktail which is necessary to start life's organic chemistry: without those six basic elements (carbon C, hydrogen H, nitrogen N, oxygen O, phosphorus P and sulphur S), you

G.F. Bignami, *We are the Martians*,
DOI 10.1007/978-88-470-2466-3_2, © Springer-Verlag Italia 2012

cannot even begin. But, of course, it takes much more, from calcium for our bones to iron for our blood, and on our Earth, there live also octopuses, crabs and scorpions, which employ copper in their (blue) blood to carry oxygen. One must also make provisions for them. In the end, to make a complete, fully inhabited planet, the more chemical elements we have the better.

This enrichment of cosmic gas by means of elements different from hydrogen and helium is called *metallicity* by astronomers (a bit of an exaggeration, but anything heavier than helium is a "metal" to astronomers/cosmologists...). Since everything in the Mendeleev table—except a dash of light cosmological nuclei—is created by stars, all in all, it is better to wait some time after Big Bang.

The first generation of stars, the one that in fact we have never yet observed, turned on in a universe very poor in complex elements and was thus obliged to take the first crucial step towards our own metallicity. As we have seen, they were great big stars, and they were short lived. Precisely because of their great mass, they would quickly burn up their fuel, that is, mostly hydrogen, and would then die an explosive death, enriching the surrounding (interstellar) gas with heavier elements.

From the second generation onwards, with an ever more complex nucleosynthesis, the metallicity of stars increases and at the same time also the one of interstellar gas increases.

But how do we know what are stars made of? The answer is that over the past century and a half, we have been practising *astronomical spectroscopy*. This means to analyse the light of a star and find in its colours the "signature" of the elements in the Mendeleev table as we see them on Earth, in our labs. This great success of astrophysics was a burning defeat for "armchair philosophers", such as the positivist Auguste Comte, who in 1835 would write about celestial bodies that never, no way, we would be able to study their chemical composition, nor their mineralogical composition nor finally the nature of organic beings that live on their surface. Thirty years later, the analysis of the spectre of the first star, the Sun, would arrive, and with it the chemical contents of our star were made clear.

Thus, to choose the right parent cloud, a wise star might as well wait a while. And our Sun wisely waited eight billion years. It let more than half the history of the universe go by before being born, five billion years ago, out of a cloud properly enriched with "metals".

But what is the *interstellar medium* made of, from which our parent cloud comes? Astronomers divide it into four components, which correspond to different aggregation states:

(a) A thin and hot (between ten thousand and one million degrees) component, uniformly distributed in space, made up mainly of ionized hydrogen, with a very low density: less than one particle per cubic centimetre.

(b) A second diffuse component, neutral, cooler (some $100°K$) and thicker (up to 100 neutral H atoms per cm^3), but still transparent to stellar ultraviolet radiation. Here molecules can begin to exist, but only very simple ones: UV radiation would destroy the more complex ones.

(c) A clumpy component, collapsed into small clouds and cloudlets, the most frequent case, distributed in a more or less uniform way throughout the Galaxy, each with a mass in the range of a few tens of solar masses. These clouds are also known as *molecular clouds*, because their density allows the formation and survival of hydrogen in molecular form (up to 100 molecules of H_2 per cm^3). It is from the collapse of this sort of clouds that stars like the Sun are born.

(d) And finally, giant molecular clouds, mainly distributed in the spiral arms of the Galaxy, with a high density, low temperatures (20 K) and a mass reaching 10,000 solar masses. It's from these big clouds that the more massive stars are generated. Giant molecular clouds are few and far between compared to normal molecular clouds, but they are particularly interesting because in their dense and well-shielded interior, even very complex molecules can form.

Interstellar Molecules

The most frequent molecule in our universe, and thus in the interstellar medium, is that of hydrogen, H_2, It is followed by carbon monoxide, CO, and water, H_2O, surely the most abundant triatomic molecule. Right here we see a hint of the importance of water in the cosmos and also, maybe not by chance, in our own life itself.

Molecules such as CO and H_2O require atoms (carbon and oxygen) that were not generated in the first 3 min: they must be synthetized by stars. Recent observations tell us that CO (and maybe also water) was already present when the universe was only 800 million years old, that is, it had little more than one twentieth of its present age. This means that nucleosynthesis in the universe must have started early. Let us send an obliging thought to that first generation of stars which disappeared over ten billion years ago and which bequeathed us the first fundamental atomic nuclei of our world and of ourselves.

But let us go back to interstellar clouds, siblings (and partially of the same age) of the one from which we were born ourselves five billion years ago. What we are particularly interested in is big clouds that can host the birth of complex organic molecules, which are needed for life. From a chemical point of view, they are mainly made up of molecular hydrogen, atomic hydrogen and a small percentage of helium (which is inert), with a very important 1–2% of heavier elements.

The temperature of giant molecular clouds is some tens of degrees above absolute zero, that is, too cold for us mere mortals, but ideal for a process of molecular formation to carry on unchallenged. Their density goes from one thousand to one million particles per cm^3; they are very thin compared to our terrestrial atmosphere, of course, but are very thick by interstellar standards.

The hunt for the different sorts of molecules in these clouds began many decades ago, when radio astronomy revealed them through the detection of "molecular lines", electromagnetic emissions that are characteristic because they are uniquely defined by the structure and composition of every type of molecules. If one observes a cloud to emit the same radiation produced in a lab by a well-known molecule, one will have evidence that that molecule exists also up there, among the stars.

This sort of observation became more fruitful with space astronomy, that is, astronomy based on telescopes put in orbit around our Earth. Many wavelengths, such as the infrared, which is crucial for some molecular species, with difficulty get to cross the Earth's atmosphere.

A high number of molecular species were thus discovered in interstellar clouds. The growing molecular complexity shows that, among the stars, a real organic chemistry lab is at work. It features cosmic dimensions, all the ranges of temperatures that a chemist might wish for, a practically infinite availability of energy and aeons at your disposal as working time. We should not be surprised that such a lab could obtain interesting patents in the field of molecular synthesis.

Up to the present, more than 120 molecular species have been discovered in the sky, from CO up to molecules with 15–20 atoms. These molecules go from ammonia, NH_3, to hydrocyanic acid, HCN; from acetic acid, $CH_3CO\text{-}OH$, to diethyl ether, $CH_3CH_2OCH_2CH_3$ and so on, ever more complex. Of course, such complex molecules only survive inside dense molecular clouds, where they are shielded from the stars' ultraviolet radiation. As we have seen, complex molecules don't like to acquire a tan: they are held together by rather weak chemical bonds, and UV radiation has energy enough to destroy them.

Life Bricks and Stardust

From one complexity to the next, in the right interstellar clouds, amino acids are born too, the "bricks of life": the ones that in 20 different species make proteins and thus make up ourselves as well. Exactly how they are formed, no one yet knows. Maybe one can get an idea of the possible chemical way of formation for the simplest biological amino acid, glycine NH_2CH_2COOH. It could rise from the occasional encounter (but in the right conditions) between such organic molecules as acetic acid with nitrogen-based molecules, such as ammonia.

Interstellar amino acids are difficult molecules to observe, because they hide inside thick and opaque clouds, where UV rays will not destroy them. Another reason is that, even with the best radio telescopes, one gets lost in the "forest" of molecular lines and bands generated by the different species of complex molecules.

Luckily, it's not only observation from a distance that reveals to us the existence in the sky of the bricks of life. Amino acids themselves, as we shall see in the coming chapters, come to see us, riding on meteorites. Even better, we will see ourselves go pluck them from the tails of comets and bring them home.

But molecules on their own, even complex ones, are not enough to make up a world in which to live. In between stars, it takes something more solid and more massive, which those molecules can cling to, in order to survive more easily and maybe further increase their complexity.

Between star and star, besides gas in the various stages of aggregation that we've been seeing, also grains of interstellar dust are to be found. They start really tiny, with sizes from one tenth to a few tens of micrometres, and are made of a refractory matter. For instance, they have a nucleus of amorphous silicon surrounded by carbon, covered on the outside by water or carbon or methane ices. Some of them are even carbon crystals, tiny diamonds a thousandth of a millimetre in size...most pure, but a bit small for a ring or an earring.

What is the origin of this precious stardust that will lead, little by little, to much bigger solid bodies building up in the sky? Its grains could have formed directly in the expanding atmospheres of certain special stars or in the immense shock generated in stellar as well as in interstellar matter by the explosion of a supernova: we don't really know. However, grains luckily exist, and they move in the interstellar medium pushed by the pressure of interstellar radiation. When they clash into each other, they may break, but sometimes on the contrary, they stick, or melt into each other, becoming bigger and bigger centres of further condensation/aggregation into big and small molecular clouds.

From Clouds to Nests of Stars

Stars are almost never born alone: often they form a system, that is to say, a nestful of stars, all belonging to the same generation.

The formation of a stellar system begins with the collapse of the denser parts of a molecular cloud: exactly those internal parts, screened from hurtful radiation, that contain the most complex and interesting molecules as well as grains of interstellar dust. As we have already seen, "collapse" simply means that the matter of which a cloud is made begins falling on itself, that is, towards its own centre. This phenomenon becomes inevitable when the mass of the cloud increases beyond a certain limit, but the collapse is very much fostered if, for example, a supernova explodes in the vicinity: that is, a giant star belonging to a former generation dies in a spectacular way.

The great bang of the supernova, which in a single second emits more energy than the Sun emitted and will emit in its whole life, propagates with a shock wave that shakes gas and clouds along its way. The compression created by the wave can generate an increase in local density which is sufficient to start the collapse of the already ripe interstellar clouds, if they are still a bit unresolved. . . . For this reason too, stars tend to be born in clusters, even by the thousand, exactly where the stars of the former generations are dying, those that have enriched the cloud with precious metals.

The beginning of the collapse thickens the central regions of the cloud, and the compression thus generates an increase in temperature. Moreover, the cloud's gaseous matter, still without stars, as it condenses sees its natural speed increase. Interstellar clouds never keep steady; they could not: they slowly and majestically turn upon themselves, like everything in the universe.

As it turns and turns, the collapsing system organizes itself into a disc-shaped structure, denser and denser, smaller in size, of course, than an interstellar cloud, but yet much bigger than a Solar System like ours. Discs still being formed or that have just taken shape are easily to be seen through telescopes' eyes; for instance, in the great Orion nebulous, which is a very active seat of stellar formation (Fig. 2.1).

They are observable as dark and flattened ellipses: for, exactly because of their high density, they absorb the light which is emitted behind them and even possibly from inside them. These *protostellar discs*, the outcome of the collapse of the cloud of gas and dust, sometimes show at their centre a luminous point: it is a sign that *there* density and temperature have increased to the point that it allows the collapsed gas to start the famous thermonuclear fusion reactions. When the energy in this reaction, which is violently exo-thermic, arrives as far as the surface of the sphere of gas and makes it shine, a star is born.

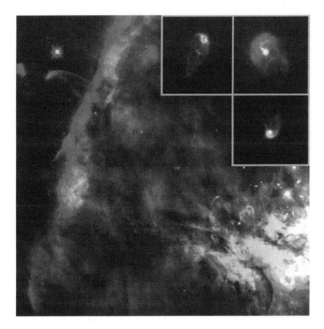

Fig. 2.1 The Orion nebula and three protostellar discs as observed inside it by Hubble telescope. Photo: NASA/ESA, M. Robberto (STScI and Hubble Space Telescope Orion Treasury Project), J. Bally (University of Colorado, Boulder), H. Throop (Southern Research Institute, Boulder), C O'Dell (Vanderbilt University)

The duration and the characteristics of the life of a star are determined in a quite exact way by the mass with which it is born. Sky observation tells us that rising stars can have a mass varying by a 1,000 factor, between one tenth and one hundred times the mass of the Sun.

What determines the mass of a star? Its physics, of course. And the real prodigy is that we humans have managed to understand in which way: we have known it for the past half century, or little more: nothing if compared with thousands of years during which men have been looking at stars and less than nothing if compared to the life of the universe. Having discovered the laws that describe electromagnetic interactions, both gravitational and nuclear, we can thus compute the minimum and maximum mass of stars.

The thinking goes something like this: to get nuclear fusion started, two protons have to almost come in collide surmounting the repulsion due to their equal-sign electric charge; for, only then nuclear forces, which are far more intense than the electrostatic ones, but only act from very close, will oblige the two particles to fuse.

When the process is repeated, involving four particles, a nucleus of helium is formed; this has a mass slightly smaller than its components, and the rest is liberated as energy according to the relation $E = mc^2$; thus is the star turned on.

But how can protons get close to one another? They must move quickly; that is, it is necessary that the gas of which they are a part be very hot, and for this, it is necessary that the mass of the gas that contracts itself be above a certain temperature. When the reckoning is done, it is found that from the cloud, in order for the fusion to take place, a mass of at least 10^{29} kilos of hydrogen must collapse; this is about a tenth of the Sun's mass. That is why stars as big as an orange cannot exist.

But why is there an upper limit to mass? This too is determined by physics. The gas in the cloud keeps falling and making the star ever bigger, until the energy it radiates is so intense that, by pressing on the falling gas, it stops it from falling. This event marks the end of the initial growth of the star and thus sets the limit to its mass.

When the reckoning is done, the mass of the stars is found to reach a maximum around 10^{32} kg, equalling a few hundred solar masses. We should be thankful it is so: otherwise by now, there would be one single gigantic big star, which would soon end by becoming a huge black hole, with a premature end for the whole universe. Luckily the laws of physics exist!

Fasting Stars

What determines the length of a star's life? The fact that it simply ends by starving to death. At its birth, it already has all the fuel it will have at its disposal to live: it is the gas of its originating cloud, made up of light nuclei, that slowly the star "digests", changing them into heavier nuclei that thus go to fill up our Mendeleev table.

During its life, a star does not eat: it must have enough with what it originally was endowed with from the cloud. But its "metabolism", which is the speed at which it uses up its nuclear fuel, is determined by its initial mass. Heavier stars are more efficient in compressing the gas inside them; therefore, they make the reactions of fusion go faster, burning a lot of fuel and emitting a lot of energy. Exactly because they are so big, they last a short time, as short as a mere ten million years.

More slender stars, such as the Sun that wears a size M, have a slower metabolism, emit less energy than the XL (luckily, otherwise, instead of tanned, we would end up toasted) and live much longer, up to ten billion years, exactly as the Sun will do, all the time going without eating, simply by making the fuel received at birth last longer.

So stars are born with an initial mass that can vary also by a thousand factors, and the duration of their lives has the same variability, 1,000 factor.

A comparison with human beings would here be inappropriate, but at the same time amusing. For us too, the variability of mass at birth is similar to that of life expectancy, but with a factor much smaller than the stars': it is not worth 1,000 but about two. We are in fact born weighing something between 2 and 4 kilos (there are no 2,000 kg babies), and we live between 60 and 120 years (to be clear, nobody lives 60,000 years). A curious coincidence.

What if human beings, like stars, were to live without eating? Let us suppose that animal fat were the most efficient fuel for a metabolism like ours. From 1 g of fat around 10 cal can be liberated. If we need 2,500 kcal a day, 1 k of fat will last 4 days, and for a year, we will need 100 k. So, in order to live 80 years without eating, we should have a "provision" at birth of 8 tons of pure fat. . . .. No, stars are much more efficient, because they don't employ chemistry, but rather thermonuclear fusion.

Let's Make Planetary Systems

When an interstellar cloud collapses, the disc at whose centre fusion will be turned on contains other interesting ingredients whose purpose is not directly to make stars; they are, however, very important for us humans.

They are the interstellar grains, of which we already spoke, that make up the first, essential step for the formation of planets. Upon this, by now, no astronomer has doubts, although we do not know in detail what is the way from grains of dust to planets. And it is most of all since we began discovering planetary systems around other stars that we understood we haven't yet understood.

Speaking of the formation of planetary systems, two elegant laws exist, simple and well proven, up to rivalling Kepler's glorious ones: we owe them to Scott Tremaine, a famous planetologist from Princeton University. The first Tremaine law says that "all theoretical anticipations on the properties of planetary systems are wrong". The second, which has an even more general character, maintains that "the safest anticipation on the formation of planets is that it is impossible".

Planets on the contrary exist: it's a fact. From the point of view of theoretical comprehension, the first to see the matter right was Immanuel Kant in 1755. Kant at the time was 31, but he already boded well: he published a small volume, of almost 300 pages, humbly entitled *Naturgeschichte und Theorie des Himmels*, that is to say, "General History of Nature and Theory of the Sky". Here, in a rather vague, but certainly innovative way, from the very preface, he stated "Gebet mir Materie, ich will eine Welt daraus bauen": "Give me matter, I'll build a world with it!"

Kant's idea was that around "nebulous stars", matter would orbit endowed with an attractive force and obeying the laws of classical physics: thus, much would suffice to explain the aggregation of planets. It is the so-called *nebular hypothesis*, although Kant would not call it like that: planets are formed from the disc of matter that is left around a star after this has been formed from a cloud.

Today we know Kant was generally right. In the two and a half centuries that have elapsed, however, some progress has been achieved. We start exactly from gas and most of all from the grains of interstellar dust. These contain carbon in the form of graphite and diamond; they are often covered in various types of ice and also contain aluminium oxides, silicon carbides and other interesting material: the famous "metals" for which it was worth waiting a few generations of stars from the beginning of the universe.

Grains too, like the gas that makes up the central star, collapse towards the centre of the cloud and arrange themselves, just because they are heavy, on the equatorial plan of the rotating disc. This transition phase from cloud to disc for our Solar System took place

Fig. 2.2 Artist's view of a protoplanetary disc. In the central part, planets under formation are sweeping away the disc's dust. Source: NASA/JPL-Caltech

roughly five billion years ago, and it was very quick (by a cosmic time scale): it only took one million years.

Once close to each other, within the disc, grains begin clinging to each other and quickly become objects measuring around 1 m; how and why this happens is not known precisely, but somehow it must have happened. Mere gravity is not enough to explain it; a help to explaining this (fact) comes from the fact that aggregating particles lie on the disc in contiguous orbits, thus moving at relative limited speeds.

Similar considerations hold true for the following phase, when boulders measuring 1 m unite to form huge rocks with a diameter measuring around 1 km, the so-called planetisms; this phase too is not yet well understood.

From this point onwards, however, the force of gravity is sufficient to explain how very quickly (a few tens of thousands of years) planetisms measuring a few kilometres move into the next stage of planetary embryos, measuring around 1,000 km. Gravity now imposes its symmetry and makes cosmic objects take on a spherical shape (Fig. 2.2).

The final process of the formation of planets comes to an end with disconcerting speed: the planetary disc dissolves in only seven million years. We found it out by observing protoplanetary discs discovered around stars close to the Sun; after that period, significant amounts of dust were no longer seen.

Summarizing, since the beginning of the process of collapse of the gas-and-dust mother cloud to the turning on of the star and then to the formation of the planetary system around it, only about ten million years go by. It is a very short time on a cosmic scale: equal, for instance, to a single thousandth part of the life of the Sun or of the Earth (but for us too gestation—pregnancy only lasts 100th of a lifetime). This result, which Kant could not know, is certainly very important: it imposed itself with the observation of the outer planetary systems, as well as with the measurement of the age of the oldest objects in our Solar System. And yet Tremaine is right: there is no theory that can explain how a thing so complicated as a planet can be made so quickly, something so different from a gas cloud (that has a density laughingly lower) and also from a star (that is made of very hot plasma). According to theoretical astronomy, in the end, the Solar System should not exist.

Let's Make Our Solar System

Theory and its models, however, do tell us something good: a great success, for instance, is the correct foretelling of the "line of perennial snow" in the Solar System. The reasoning moves from the *temperature profile* of the protoplanetary disc, that is, of the way in which its temperature varies from centre to periphery. This characteristic has particularly important consequences for the destiny of waters.

In space, H_2O goes directly from the solid to the gaseous state, and this *sublimation* takes place at about 120° below Celsius zero (i.e. at about 150 K). Computing the temperature of the just-formed disc in relation of the distance from the star, one finds that water remains in a solid state (ice) only beyond a certain distance equalling five times the distance Earth–Sun. As on the mountains, this height above the sun is called *snow line*.

Beyond the orbit of Mars, as it were, solid ice could exist from the beginning; also in great quantity, this side it could not. Earth, for instance, was born in a zone where temperature measured many 100° and ice could not exist. In all the internal part of the Solar System, where the four rocky planets formed—Mercury, Venus, Earth and Mars—solid and dense, at the beginning, there was very little water: it was too hot.

In the outer part, on the contrary, beyond the snow line, there was a lot of ice in the form of comets. But there were also, in formation, the four gaseous giants: Jove, Saturn, Uranus and Neptune. We call them gaseous, and with reason, these four outer giants, but we know they have a solid core, although with a low specific weight. It is by now an acquired fact that first the core was formed and then, because of gravity due to its mass, gas thickened around it.

It has been calculated that the mass of the core of the outer giants must equal around 20 times the mass of the Earth for it to be able to hold the thick blanket of gas that covers them. So, the gaseous giants, although massive, are not very dense, because the great distance from the Sun has allowed them to hold water, hydrogen and helium in a solid form; they rather grew exactly by capturing these light elements that lay in their vicinity.

Beyond Neptune, on the contrary, there is a great quantity of bodies of small and medium dimensions, either rocky or frozen, which escaped the gravity of the giant planets. An example of this is Pluto, which for half a century was considered a planet and was then downgraded to "nanoplanet" like many others. But there are billions of comets, small celestial bodies made up of solid water.

These outer objects are very important for our research: for, they did not undergo the heating process of the inside part of the Solar System when being formed and have thus remained "uncontaminated": they kept intact, for instance, the big organic molecules which, as we have been seeing, had already formed in the mother cloud.

The huge gravitational fields of the outer giant planets tend to attract and release the small bodies orbiting outside them, be they asteroids or comets. The result is that the latter can be expelled from the Solar System and start travelling towards other stars and planetary systems (and certainly also a symmetrical phenomenon takes place

Fig. 2.3 Artist's view of the cosmic bombardment undergone by Earth during the so-called age of Hades. Image Copyright Take 27 Ltd., courtesy of Julian Baum

with bodies coming to us from other systems); they could also be attracted to the inner part of the Solar System and get closer to the small rocky planets and maybe happen to fall exactly on the third planet from the Sun.

Here is another great success of theoretical astronomy: the explanation of the immediately following stage of planetary formation. It is reckoned that for some 400 million years, planets have been subjected to incessant bombardment. For, with their gravity, they would attract the bits of the Solar System which had remained loose, rocks (also huge ones, such as planetisms or planetary embryos) as well as comets. Thus, also our own Earth, when new born, went through a hellish time, appropriately called Hades age, that lasted exactly 400 million years, as anticipated by theory (Fig. 2.3).

Let's Make the Earth

When was the Earth formed? Today we know it precisely, thanks to dating techniques based upon various radioactive isotopes: Earth was born 4.67 ± 0.02 billion years ago. This is the time in which superficial rocks of the primeval magma became cool enough to crystallize, turning to a large extent into the solid state. The intense bombardment of the age of Hades has then continued to heat the Earth (and its rocky colleagues) intensely and to model it by means of catastrophic events.

Let's take, for example, the formation of the Moon: it took place some 4.52 billion years ago when an object as big as Mars came to clash against Earth, getting fused by the impact and contributing to the formation of the present terrestrial nucleus. But this is not all: the clash detached a bit of this mix of Earth and of the unknown "impactor", thus generating our satellite. That is why the Moon has a composition slightly different from the Earth's: it still carries traces of pieces of a foreign body.

Similar catastrophes have probably been the lot of Venus too, which maybe for this reason even changed the sense of its rotation round its own axis, and to Mercury, whose rocky cloak (mantle) has been completely destroyed.

But let's go back to the Earth when being formed, to our ball of cooling lava, still without both water and atmosphere. The first solid continental masses were formed 4.42 billion years ago: it is the age of some zircon crystals found in Western Australia. By common agreement, it has been decided that the "formation" of the Earth ends at that date, although some external bodies have kept falling on the planet (and still at the present time, every year Earth receives tens of thousands of tons of celestial material—from the sky).

The mass of the Earth, although not huge, has allowed it to rapidly capture a bit of gas, to form a primitive atmosphere made up of aqueous vapour, carbon dioxide, nitrogen and hydrogen: the perfect mix for greenhouse gas. And thus the primitive atmosphere, just because of the greenhouse effect, must have reached a temperature of 2,000°C with a huge pressure, something like 1,000 atmospheres: not a very pleasant spot.

Exactly such a high temperature, however, would produce a very high infrared emission in the direction of outer space, quickly cooling the planet's surface until this allowed the existence of liquid water; this would start very hot torrential rains. Thus, oceans were formed, which also profited from the plentiful extraterrestrial shower of comets. Carbon dioxide dissolved in water forming carbonic acid, which reacted with the existing minerals to form stable salts (such as calcium carbonate and silicon oxides) and forever removing most CO_2 from the atmosphere.

Little by little Earth cooled down and things improved: there were great masses of liquid water, and, with the decrease of CO_2 in the atmosphere, the greenhouse effect was growing more bearable. The second phase could thus begin, the one of organic chemistry. Thanks to nucleosynthesis (both primeval and stellar) and to successive generations of stellar collapses, this new planet had inherited the magic cocktail CHNOPS, although in a rather strange and diluted form, and it was also endowed with many more chemical elements, all the way to copper for the scorpions' blue blood.

But the abundance of free oxygen, which we know today, was not there yet. Oxygen is a very aggressive gas, which combines chemically with great facility, as when it joins with iron generating rust. If one wants a lot of free oxygen to be available, therefore, one must keep producing it, and on the Earth about four billion years ago, it became widespread, because life started, that curious phenomenon that could exactly produce free oxygen.

Chapter 3
Astronomy in Search of Alien Planets

As it will often happen, it is difficult to say who really was the first who thought up this matter. The existence of planets outside the solar system had already been posited, at least as a philosophical hypothesis, more than 2,000 years ago.

Epicurus had said (in Greek): 'There are infinite worlds, both similar to ours and different from it'. This idea of his, of extraordinary modernity, however, did not convince Aristotle, who in the field of astronomy had very conservative ideas.

Later on, over the centuries, many picked it up again, and in 1600, Giordano Bruno suffered at the stake as an impenitent heretic, among other reasons, for saying in 1584 that 'numberless suns exist, numberless earths rotate around them [. . .] These worlds are inhabited by living beings'; he would repeat these utterances again and again, even under duress.

It is a fascinating notion, and in 1678 it moved Christian Huygens (the Dutch astronomer who discovered Titanus, the biggest of Jupiter's satellites) to try and reveal planets around other stars.

But observations proved too difficult even for those that at the time were the newest telescopes of the Observatoire de Paris, where Huygens was working with Giovanni Domenico Cassini (Jean-Dominique Cassini).

Around 1830, Giacomo Leopardi, who was fascinated by astronomy, in his *Operette morali* (Essay and Dialogues) has Copernicus say in a dialogue with the Sun that thought that every little star had its own planet.

G.F. Bignami, *We are the Martians*,
DOI 10.1007/978-88-470-2466-3_3, © Springer-Verlag Italia 2012

So, it was thousands of years that the matter had been thought about, the idea was in the air, but, until a few years ago, nobody had been able to identify one planet outside the solar system.

For astronomers, directly 'seeing' a planet which orbits around another star, even the closest to us, would appear an impossible feat, not only the object to be identified is tiny, but the light emitted by the star is millions of times more intense than the one diffused by the planet. Think how difficult it is to tell what colour a traffic light is when the Sun is directly behind it; well, the problem here is much harder.

The resourcefulness of researchers and the more and more sophisticated technology of instruments of observation have been able, in recent years, to develop methods which, albeit mostly indirectly, allow us to demonstrate with certainty the existence of planets around other stars and even to measure their characteristics.

The First Extrasolar Planet

It would be beautiful if we could tell that it all happened in one fatal night, as the case was for Galileo and his observation of Jupiter's satellites.

On the contrary, the discovery of the first extrasolar planet, which dates back to 1995, required of two Swiss astronomers, Geneva Professor Michel Mayor and his student Didier Queloz, many nights of observation, to be followed by an analysis of the data that lasted many months.

At the Observatoire de Haute-Provence, in the South of France, Mayor and Queloz were under pressure because they were working in direct competition with the great Lick Observatory in California. They had concentrated on star 51 Peg, number 51 in the Pegasus constellation. They had chosen it because it is close enough, only 40 light years away from us, and thus shining bright and easy to observe; after all, every great discovery requires a bit of luck.

Mayor and Queloz would frequently measure the light spectre of the star, knowing that it would undergo a regular change if the star had a periodical repetitive movement.

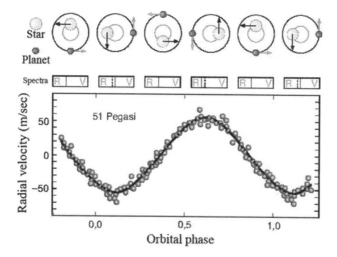

Fig. 3.1 When a planet and its star orbit around their centre of mass, spectral stripes of stellar light move according to Doppler effect towards either the *red* or the *violet extremes* of the spectre. From their shifting, one can deduce the radial speed of the star. The experimental data of the graph are those published in the article by Mayor and Queloz that announced the first discovery of an extrasolar planet (*Nature* 378: 355–359, 1995)

When there's a planet orbiting around a star, both move—according to Newton's laws—by rotating around the mass centre of the system. If the planet is massive enough, the effect is meaningful, and it can be remarked that the star periodically comes closer and goes further away (Fig. 3.1).

From the spectre of the star under observation, one can infer its *radial speed*, that is to say, the one that follows the direction of our line of sight: because of the Doppler effect—similar to the one that makes the pitch of the whistle of a train coming close and then moving away change—the spectral stripes of its light move towards the red, when the star is moving away from Earth, or towards the blue, when the star is coming closer.

Unremittingly, the two Swiss measured time and again the spectre of 51 Peg by applying this *method of radial speeds*. They were convinced that it would take a good many years in order to find something interesting. But on their instrument, they had also placed a little box containing electronic devices that would every minute check the stability of the whole system of measurements. And the little box would indicate a disquieting variableness of data day by day: it was the opposite of the long-term stability which astronomers were expecting. Believing in a malfunctioning, they checked and checked again everything more than once.

Little by little, however, Mayor and Queloz were convinced of something that seemed incredible, exactly thanks to the little box that the American rivals did not have: they were the first to discover a 'Jupiter' that only takes 4 days to revolve around its own sun! They went to Florence where an international astronomy conference was taking place and there presented their results, which are shown in Fig. 3.1. Thus, in 1995, a turning point took place in astronomy and also, a little bit, for humankind.

Since the influence of planets on the motion of their star is so much greater the bigger they are, it is not surprising that the planet orbiting around 51 Peg be a giant with a mass comparable to Jupiter's. What surprises, on the contrary, is its very short orbital period which, as Kepler taught us, indicates a remarkable closeness to the star: the planet of 51 Peg is a gas giant much closer to its star than our tiny Mercury is close to the Sun. It is a hot 'Jupiter', indeed a very hot one, reaching 2,000°C, much different from our own freezing Jupiter at −200°C. But how can it exist? Why does it not evaporate?

Here we can see right away that the first planet discovered rotating around another star lets us understand that we have to correct our general ideas on planets. Up to this point, such ideas were simply based on our solar system, and nowadays we even surmise that maybe we have not understood much of this too. ...

Today, 20 years on, we are about to reach the goal of 1000 extrasolar planets already identified. The measurement techniques have been refined, cutting down more and more the limits of the mass of identifiable planets: thanks to high-resolution spectroscopy, the method of radial speeds allows the identification of planets whose mass is smaller than Neptune's, even down to few terrestrial masses. These planets, however, must be close to their star: in fact to be sure that the perturbing is real, it is always necessary to observe several cycles, but faraway planets have a longer period of revolution, thus to observe its effects, years and years of observation may be necessary.

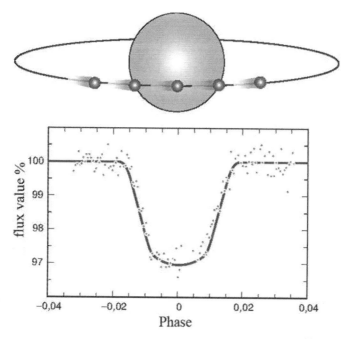

Fig. 3.2 When a planet, seen from the Earth, passes in front of its star, the star's luminosity lowers down. These experimental data have allowed François Bouchy and other astronomers, among them (whom)Mayor and Queloz, to discover in 2005 a gaseous giant orbiting around HD189733 star of the Vulpecula constellation, 63 light years away from the Sun

The Transits Method

Meanwhile, other techniques of observation have been developed. The *transits method* is based on photometry, that is, on a count of photons, a very precise measurement of the flux of light coming from a star. This technique allows the reduced luminosity of a star to be perceived when the small disc of a planet passes in front of it (Fig. 3.2). It is the phenomenon in the solar system that we can observe when, twice every century, Venus passes in front of the solar disc.

The observation requires very exact and stable photometric measurements: in fact, the shadow projected by the planet only covers a very small part of the star's surface; thus, variations in luminosity to be measured are tiny (usually much smaller if compared to the 3% which is visible in the example of Fig. 3.2).

The transits method also requires a great amount of luck: it is necessary that, from the point of view of Earth, the planet should pass in front of its star's disc, which seldom happens.

So the hope of identifying the 'right' star with a 'right' planet one must watch a lot of stars and discard all the occasional variations that a star can have without this indicating the transit of a planet, a lot of work, which requires patience and care.

The first example of a transit was observed in 1999 about another star in the Pegasus constellation, HD209458, chosen because the technique of radial speeds had already signalled the presence of a planet around the star.

While the method of radial speeds supplies information on the mass of the planet and on its orbit, the transit method allows us to estimate the radius of the planet. The combination of the two methods thus allows us to arrive at a much more complete characterization: knowing both mass and radius of the planet, one may determine its density, thus establishing if it is the case of a gas or of a rocky planet.

As we have seen, for the transits method to be effective, it is necessary that a great number of stars be kept under observation over long periods of time, while keeping up measurements of the luminous flux of each with great precision. Doing such a thing from the ground is very difficult, mainly because of the variableness of atmospheric conditions; for the purpose of this type of observation, it would be ideal to employ telescopes set in space, outside the atmosphere.

Two space missions are at present in search of extrasolar planets by the transits method: COROT, a small scientific mission launched at the end of 2006, in cooperation between France and European Space Agency (ESA) and Kepler, a NASA mission which has been orbiting since March 2009. COROT, in roughly 3 years, had discovered nine planets, while Kepler is discovering them by the thousand, of which about 150 possible 'new Earths', that is, planets with a radius up to twice the terrestrial one. Who knows that among these there may be the right one.

The Method of Gravitational Lenses

Another indirect method for the search after extrasolar planets exploits the huge databases which have been accumulated for the study of the effect called *gravitational lens*, a consequence of Einstein's theory of general relativity.

The photons that make up the luminous radiation possess energy: because of this, they feel gravitational attraction as if they were particles with a mass. The light emitted by a star is thus deviated if, before reaching us, it passes by a massive object that, by chance, happens to be on a line with the star and the Earth. The result of the deviation can be, like in a lens, the focalization of the star's light.

It is thus possible to reveal the presence of an extrasolar planet when it, by chance, is in line in a very precise way with its mother star and generates an effect of gravitational 'microlens'. For this type of measurement, wide-field telescopes are used, based on the ground, and one looks towards the centre of our galaxy, where billions of stars are potential candidates.

Observing them with great care and continuity, it can happen that a sudden increase in a star's luminosity can be suddenly glimpsed: if we can exclude that the phenomenon is due to the regular variations intrinsic in stellar light, that immediately means that indirectly the presence of a planet has been identified.

This method of *microlensing* has the advantage of allowing the discovery of planets around faraway stars. In the method of the radial velocities as well as the transits, one must concentrate on very brilliant stars that are therefore close by; microlens events, on the contrary, can concern stars many thousands of light years away from us.

There is therefore a possibility of collecting samples from very wide stellar zones and to check if the presence of planets is to be commonly found also in faraway parts of the galaxy. Moreover, microlensing is potentially sensitive also to planets with a small mass, such as the Earth, that are on the contrary much more difficult to be observed by means of the other two methods.

Other Methods Employed in 'Seeing' Extrasolar Planets

Very soon we'll be able to reveal extrasolar planets also by means of a fourth method, based on the measurement of the shifts that their perturbing presence causes in the star around which they revolve (rotate). Since we are talking about very tiny shifts, it takes instruments of the greatest angular resolution set in space, outside the terrestrial atmosphere.

The most promising space mission is GAIA, of the ESA, that continues the great European tradition in the exact positioning of stars, which began 20 years ago with mission Hypparcos. GAIA will be able to determine, with the highest precision, the position of hundreds of millions of stars and thus to measure their movements as projected on the celestial sky. As a by-product of this immense work of analysis of stellar and galactic dynamics, GAIA will offer us thousands, even hundreds of thousands of new planets. With all these potential targets, it will be possible to proceed with a research aimed at those planets most interesting for us by making use of those space technologies that are even more refined and that we are already inventing.

Finally, though this can be very difficult, it is also possible to reveal in a direct way extrasolar planets by taking photographs of them. This requires instruments with a very high angular resolution, which should go together with the use of the very large telescope (VLT) with strategies aimed at excluding the light of the star from the signal. Some planets have already been filmed in this way, by making use of ESO's VLT, Europe's austral observatory located in Chile.

They are in any case rather massive planets, definitely far away from their star. In the case of Fig. 3.3, the planet photographed by Hubble Space Telescope is so far removed from its star that it takes more than 800 years to complete an orbit ... how cold must it be on that planet!

Here we are completely at the other end from the case of the planets found to be very close to their stars, with a year that can last just a few days. We have thus discovered that extrasolar planets, besides being very numerous, belong to all kinds.

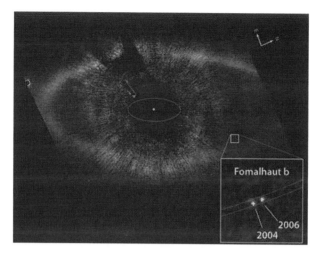

Fig. 3.3 The Fomalhaut star, obscured in order to take this picture by means of Hubble space telescope, is surrounded by a disc of dust in which the first extrasolar planet ever to be photographed is to be found. The oval tracing in the centre indicates the dimensions of our system: the Fomalhaut b planet is thus very far away from its star, and, as years go by, it moves very slowly along its own orbit. Photo NASA/ESA, P. Kalas, J. Graham, E. Chiang, E. Kite (University of California, Berkeley), M. Clampin (Goddard Space Flight Center), M. Fitzgerald (Lawrence Livermore National Laboratory) and K. Stapelfeld and J. Krist (Jet Propulsion Laboratory)

What Can We Learn from the Planets We Already Found?

At the Web address http://www.exoplanet.eu/catalog.php, one can find an up-to-date listing, in constant and quick evolution, of extrasolar planets observed up to the present. Let us see now the characteristics of these planets and of their stars.

The planets' masses range from between a few terrestrial masses and a few tens of Jupiterian masses. The planet–star distance in most cases is only a fraction of an astronomic unit (1 AU is the average Earth–Sun distance, equalling some 150 million kilometres), and the corresponding orbital period is around 3 days.

They are indeed planets resembling the gas giants of our solar system, which however orbit around their suns at a distance lower than Mercury's. This is an important result, not least because it is completely unforeseen: for stars similar to the Sun, it was expected we would find planets similar to Jupiter at a distance of 4 or 5 AU, not simply at 0.05 AU from their star.

Moreover, many of the planets found move along very eccentric orbits, that is to say, not circular, but rather very elliptic ones. It is a phenomenon not to be found in our solar system, and it is with difficulty reconciled with the idea that planets are formed inside interplanetary discs: viscosity due to dust that makes up the disc should in fact rapidly lead to almost circular orbits.

Many among the planetary systems found to date contain more than one planet, but this could come from a limit in the method of radial speeds, until the present the most frequently employed, which measures more interesting results in the case of bigger planets.

Altogether, the analysis of data from observation indicates that the paradigm we had boastfully deduced from the observation of our solar system is, if not wrong, limited to say the least of it.

The theories of planetary formation, developed before 1995, were aimed at explaining the formation of small rocky planets in the inside parts of planetary systems, while the gas (and very cold) giants were to remain confined in the outer parts, further away from the star. As we saw in Chap. 2, the two principal theoretical models are fragmentation of the protoplanetary disc and the growth of planetisms on a central nucleus. In both cases, it is supposed that the profile of temperature to be found in the protoplanetary disc concentrates denser material in the bodies closer to the stars, while volatile materials are confined to further away planets.

According to these theories, it should not be possible for giant planets to exist at such small distance from a star: the existence of the so-called hot Jupiters is incompatible with the high temperatures of the region where the planet should form.

After the discovery of hundreds of extrasolar planets with unexpected characteristics, it is clear that classical theories must be brought up to date.

One of the solutions suggested for this enigma is the model called *migration*, according to which the gas giants form in the most external regions, later to migrate towards inside ones. The existence of hot Jupiters demands the idea that the present position of a planet may have nothing to do with its position at birth.

The migration may have different causes: following gravitational interactions among a number of giant planets, for instance, one of the planets could have been expelled from the system, while the others moved close to the star; or else, a planet, if it has a sufficient mass, could be able to find its way towards the inside of the gas disc (which is in any case the basic structure of the formation of any planetary system).

Models of migration require that planets form very rapidly, otherwise the stellar wind might dissipate the disc before it can interact with the planet. It is not enough, however, to find a way of sparking off migration: it is also important that we identify one or more mechanisms that can stop it, thus avoiding the planet falling on its star. Yet nobody has yet been able to explain the great number of planets with an orbital period of 3 days and the paucity of those with a shorter orbital period. It could also be possible that planets that move too close to their stars quickly evaporate, thus becoming too small to be revealed.

Another interesting result that surfaces from a statistical analysis of extrasolar planets found to date is the importance of the composition of the star: the more it is rich in heavy elements, the higher its probability of having planets. Yet the composition of the star must reflect the surrounding disc's, which originated from the same cloud; it is reasonable to think that a protoplanetary disc rich in heavy elements and in dust may favour the formation of planets. Taking this as our starting point and according to the data obtained by mission Kepler in its first year of observations, planets similar to Earth could be as many as a 100 million in the Milky Way.

We are indeed discovering the fact that planetary systems present in our galaxy are very numerous, different from one another and probably almost all different from our own solar system, which makes them the more interesting.

Giordano Bruno and Giacomo Leopardi were right: around stars, planets are the rule, rather than the exception, but they are different from what we had been expecting, starting from our habitual anthropocentrism and our small imagination.

What Planets We Would Like to Find

In the coming years, especially thanks to mission GAIA, we shall identify thousands of extrasolar planets. With such a statistical basis at our disposal, we shall certainly find cases similar to Earth in mass, orbital period and distance from the central star; it will be possible to study more deeply, also by making use of new instruments with a very high resolution, such as second-generation space interferometry.

How must a planet be made to be able to accommodate forms of life? For simplicity's sake, let us limit ourselves to conditions which are necessary in order to have a biosphere similar to the terrestrial one: a rocky planet is necessary, massive enough to hold an atmosphere (Mars, for instance, has let it go) but not so massive as to hold hydrogen, which is lethal to any form of life (an atmosphere like Jupiter's would be toxic). To begin with, one can look for a planet with a mass between 0.5 and 10 terrestrial masses.

On the surface of the planet, there must be liquid water. If the pressure is of one 'regular' atmosphere, this requires a surface temperature of some 300 K (with a moderate greenhouse effect to avoid finding ourselves in Venus's infernal heat). Two more important optional requirements to be requested for sure for a planet worth living in are a planetary magnetic field to screen a possible biosphere from ionizing cosmic rays and hot spots that can carry nutrients to the surface.

In order to be positioned in the 'habitable zone', the planet must necessarily lie at a distance from its star between 0.1 and 2 AU, depending on the type of star. A distance bigger than 2 AU would require a star much hotter than the Sun, but that type of star has too short a life: it would not last the billions of years that have been necessary for life to evolve on Earth. At a distance lower than 0.1 AU, on the contrary, the temperature would be too high even with a star colder than the Sun.

The first extrasolar rocky planet with a terrestrial size, discovered in February 2009 by mission COROT, is only five times more massive than the Earth and has a density similar to the terrestrial one, but its orbit is 60 times smaller than the Earth's. The nearness to the star, which is similar to our Sun, sets the surface of the lighted (lit) hemisphere finds itself at 2,000°C, while the side which is in the shade is at $-200°C$. Clearly, this is not a planet that can accommodate life.

Among extrasolar planetary systems discovered to date, some 50% could have a habitable zone, where a rocky planet similar to ours would have the potential of accommodating the development of life. In many of those cases, however, the habitable zone could have been crossed by a migrating gas giant, which can easily destroy 'out of cannibalism' a rocky planet. In the end, the probability of finding an Earth-like planet in the habitable zone goes down to below 10%.

Yet, given the huge number of stars in our galaxy and given the frequency with which around them planetary systems are formed, in the coming years, there will certainly be no lack of planets candidate to be explored—for the time being from afar—in search of possible signs of life.

How to look for those signs? The most powerful enquiry method is spectroscopic analysis of the light reflected by the planet, a technique that however requires a space interferometer, a very expensive instrument and one still under study. The simplest spectral stripes to be searched for are oxygen's, given the fact that molecular oxygen and ozone on Earth are of biogenetic origin. Even more interesting would be the discovery of stripes of absorption, characteristic of a molecule able—like chlorophyll on Earth—to turn the energy emitted by the star into chemical energy.

Chapter 4
Contact Astronomy: The Universe Invades Us

Not only on light and other electromagnetic radiation do our study and our understanding of celestial bodies thrive. Here, we will not have recourse to gravitational waves (that would indeed allow us to get very close to Big Bang) or neutrino astronomy (that would unveil to us what happens both in supernovae and close to neutron stars), nor shall we refer to cosmic rays (that would help us understand the distribution of sources of energy in the universe).

We shall rather limit ourselves to speaking of dear old *meteorites*, or rather of all those celestial bodies, even microscopic ones, that still today hit the Earth in great numbers. Nothing, be it understood, if compared to what used to happen in the age of Ades, when the Earth's surface was overheated and made uninhabitable by the constant dropping of fragments from the proto-planetary disc which had remained floating after the formation of planets. Today, some 40,000 tons of extraterrestrial material falls on Earth every year, unavoidably a rough appraisal. Mostly, this fall concerns micrometeorites, particles measuring less than 1 mm: in other words, interplanetary dust.

This cosmic material, big and small, can be "heavy", that is, with a high percentage of minerals more or less ferrous, as is the case of Fig. 4.1, or "light", mainly composed of water, under the guise of small comets, snowballs or ice crystals. In fact, almost always, we have a mix of the rocky-metallic component and of the watery one. But on today's Earth, the watery component usually does not reach the surface: it evaporates before that because bodies that enter into the atmosphere are by friction heated up to very high temperatures.

G.F. Bignami, *We are the Martians*,
DOI 10.1007/978-88-470-2466-3_4, © Springer-Verlag Italia 2012

Fig. 4.1 The Hoba meteorite from Namibia is the most massive ever found on the surface of the Earth: 60 tons of iron and nickel

A New Science, as Old as the Hills

As far back as humankind has existed, meteorites have been observed with awe (as prodigies or signs from above), collected and adored (because they came from heaven) and also used for practical purposes. They were, for instance, a very useful source of iron for the points of spears of the Eskimos, who inhabit a frozen world devoid of mines.

More recently, we have begun to study meteorites in order to snatch the information that will take us to the birth of the solar system or maybe to zones possibly even further away in the cosmos. In general, we are here dealing with objects as old as the Earth, if not older: they formed from the condensation of the proto-planetary nebula from which the solar system originated.

Meteorites are thus a first ideal step for contact astronomy based on direct analysis, instead of analysis at a distance, of cosmic objects. In this case, we are dealing with "passive" cosmic objects because it is the universe that very kindly comes to visit us at home to get itself analysed in its components.

Apart from minerals and water, meteorites and comets carry onto Earth also a lot of organic material. Only in the first billion years, it is reckoned that on the newborn planet, 10^{22} g of carbon fell in the guise of dust; for a comparison, today's whole terrestrial biomass only contains 10^{18} g of carbon.

In bigger meteorites—that today still arrive numerous, something like 10 tons a year—a great number of organic molecules have been discovered. These substances can be easily studied in a lab, without having to engage in expensive space missions; but the price for this facilitated bioastronomy is that one must pay great attention to avoid contamination with terrestrial organic material.

Among the most interesting organic findings to be considered are some amino acids, recently found in fragments of a meteorite which fell near the city of Murchison, in Australia's Victoria state.

The Murchison Meteorite

Murchison is a carbon meteorite fallen in 1969, which probably exploded in the air before touching the ground. Its fragments (Fig. 4.2) have spread over an area of above 13 km^2, from which more than a hundred kilos of primeval material was collected.

Fig. 4.2 A fragment of the Murchison meteorite during laboratory analysis; the test tube contains molecular species extracted from this same meteorite

The scientific interest of this extraterrestrial object became clear right away because it was understood that it contained a lot of organic material, a true encyclopaedia of interplanetary chemistry (or even interstellar chemistry: some people believe it to be older than Earth and some even surmise that it can have an extrasolar origin).

Murchison contains amino acids which are quite common to us, such as alanine, glycine and glutamic acid, but also some others too, and strange ones, different from the 20 amino acids that go to make up our proteins. And the chemical analysis of these organic molecules brought with it another important surprise.

To understand what we are talking about, it is necessary to keep in mind the fact that in molecular structures atoms and bonds among them are no chance throws: often, it is exactly in structure that nature hides molecules' most interesting properties. This begins to be true of water and benzene, a much simpler one; but complex organic molecules, such as sugars and amino acids, also have interesting properties of symmetry. In particular, they often exist in two different forms, even if they are identical in structure (atoms and interatomic bonds), but are one another's mirror image, like the right and the left hands: therefore, we speak of *chirality*, from the Greek *cheir*, which means "hand".

It could be thought that, if the two mirror-image forms of a molecule—defined as laevogyrate and dextrogyrate—are identical from the point of view of complexity, which they are, then the two molecular forms should also be equally probable and thus equally present in their numbers in nature. On the contrary, on Earth, it is not so: in living beings, the "left" form, laevogyrate, of amino acids is more abundant than the dextrogyrate one (whereas for many sugars, the opposite is true). Nobody has yet understood the reason for this *omochirality*: could it be a complete terrestrial peculiarity of the "bricks of life"?

Well, the study of Murchison has shown, beyond any doubt, that amino acids that arrived from space have an asymmetry in favour of the laevogyrate form, all the way similar to the one existing on Earth. And we can stand assured it is not a matter of terrestrial contamination: the single elements, for example, nitrogen, in Murchison amino acids have an isotopic composition different from nitrogen atoms to be found in amino acids of an Australian sheep or blade of grass or in any case terrestrial. And this spectacular result, which was obtained by studying the Murchison meteorite, is confirmed by analysing the organic material in many other meteorites.

With this passive form of contact, astronomy has scored an important point, maybe impossible to be acquired by astronomy at a distance. Besides giving us their cosmogonic message, the rocks that fall from the sky have literally allowed us to taste the interplanetary soup of organic molecules in which we are steeped and from which somehow we descend.

The soup could even be interstellar, maybe galactic. By means of large telescopes, we have identified complex organic molecules even outside the solar system, in interstellar clouds that are still in our neighbourhood, but also in ones close to the centre of the Galaxy, more than 20,000 light years away.

Natural Exchanges Among Planets: It Is No Science Fiction

Incredible as it may appear, planets exchange rocks. The phenomenon has a very low, but not inexistent probability: given a sufficiently long time, thus sooner or later, it will take place. The exchange mainly concerns rocky inner planets, such as Mars, Earth (with the Moon) and Venus, but small planets as well, such as Vesta.

The qualitative mechanism is as follows: a big and speedy object, for instance, an asteroid, impacts on the surface of a planet; fragments are generated that sometimes dash away at a speed higher than the speed of escape of the planet; the fragments move within the solar system and, after a shorter or longer time, may be captured by the gravitational field of another planet and fall on it.

Of this dynamic process, quantitative models can be drawn, thus obtaining estimates of the mass transferred per time unit. In the case of exchanges between Mars and Earth, for instance, one discovers that in the last four billion years—the "useful" period for life—on to our planet and coming from Mars, many billions of rocky blocks measuring at least 1 m have fallen. Even today, in the general fall of meteorites, every year from Mars, we get some half ton of rocks, great and small!

To travel from Mars to Earth, all this material employs quite some time, from hundreds to thousands up to millions of years. Fragments don't go a straight way: they follow the rules of celestial mechanics which, although dominated by the simple law of universal gravitation, lead to complex trajectories, or rather to chaotic ones, because of the several bodies concerned.

The same considerations apply, *mutatis mutandis*, to the transfer of material from Earth to Mars. The phenomenon is slightly less probable because Earth is more massive (and therefore its speed of escape is higher) and because the terrestrial atmosphere has always been thicker than the Martian one. It is reckoned that in the past four billion years, Mars may have arrived 50 million terrestrial rocks measuring 1 m.

Such computing can also be done for exchanges with the Moon, with Venus as well as with big asteroids such as Vesta. Exchanges of planetary material are the rule in the solar system, and their quantitative volume is determined by the characteristics of the bodies involved as well as by their distance.

The Mars–Earth case, however, has been studied in better detail, mainly thinking of the possibility of an exchange of living matter between the two planets. If on Mars there has been (or there is) life, albeit in the elementary form of bacteria or unicellular eukaryotes, the question rises spontaneously: could Martian life have arrived on Earth, carried by any of the rocks arriving from Mars? Otherwise said, is it not by chance that *we are Martians too*? Or, on the contrary, has it been maybe a terrestrial rock that brought life on Mars, if ever there was, or there is, one any? *Are they maybe terrestrial*?

Such a discovery of course would not solve the problem of the origin of life: it would simply shift it somewhere else or make it more general. But it would be an interesting result.

The hypothesis of an exchange of living matter among bodies of the solar system can be studied in a rational way by subdividing it into three stages.

The first stage is the departure, for example, from Mars. It is necessary that the form of life possibly present on the planet's surface should not be destroyed in the impact that sets off the fragments on their interplanetary flight. Speaking of this, the decisive factors are temperature and acceleration.

It can be demonstrated that, even if the impacts of which we spoke disperse a great amount of power, by either melting or very much heating a large part of the fragments, there is a real possibility of survival for forms of life on the ejected pieces of rock, mainly on those that lie on the margin of the point of impact. Several fragments can be expelled, without their temperature going beyond 100°C, which we shall arbitrarily take as the upper limit for survival. It is estimated in a billion the number of fragments measuring 1 m arrived to date from Mars on Earth that, completely or in part, did not go above 100°C.

As for acceleration, there is no problem, principally if we are speaking of simple organisms. At least two stocks of terrestrial bacteria (*Bacillus subtilis* and *Deinococcus radiodurans*) live happily after being treated to terrifying spinnings by centrifugating machines or even after being shot inside high-speed bullets, thus undergoing an acceleration half a million times bigger than gravity's. For sure, not all of them can make it: it is reckoned that only one microorganism out of 10,000 could survive the impact and, still living, quit the planet. But if we posit, for instance, a density of bacterial spores amounting to 100 millions per gram of matter, which is the typical amount in a terrestrial desert, on board a kilo of rock ejected from the Martian soil there would be all of *ten millions of living spores on their way.*

The second stage is the long interplanetary flight that can last up to one million years. Space is a hostile environment: void, with extremes in temperatures and, most of all, with all sorts of ionizing radiations, from solar ultraviolet to galactic cosmic rays. But there is a remedy to everything. Void and lack of water and nourishment spores can compensate by drying up, assuming that, once the environment is back in an acceptable condition, they are then able to repair the damage operated by void and by radiation to the genetic material.

Radiation and, in part, the terrible thermal excursions those spores can survive are hidden in the inner cracks of ejected rocks. The bulkier the rock, the better will be the protection (and let's keep in mind that bacteria can be very hardy: *radiodurans*, as its name suggests, can even live inside nuclear reactors).

Of course, it is important to cut down as far as possible the exposure time to the hard conditions of space: favoured are thus those fragments that do not waste time loitering around the solar system, and statistically, there's always one.

So, if on Mars there were forms of life similar to terrestrial bacteria, the interplanetary voyage would not suffice to kill them all on all the fragments; the ideal transport would consist of the biggest and quickest fragments.

It is impossible to arrive at an exact assessment, but reckonings indicate that in the case of a million-year voyage, with the protection of 1 m of rock, one spore out of one million can survive. Keeping in mind the fact that we left Mars with ten million spores per kilogram, we are in any case left with *ten live spores* when in view of the Earth, ready to face the final hurdle.

The third stage is the entrance into the terrestrial atmosphere: here too friction can generate high temperatures, but it's negligible stuff if compared with the conditions of impact at departure. Many meteorites just fallen on terrestrial soil can be easily picked up with your hand: it is probably the inner parts of a bigger meteorite that split up after the fusion of the outer parts. At this stage too, there's an inner part of the rock that does not go above 100°C. Let's suppose this to be one tenth of the mass of the meteorite: if this is the case, *at least one spore makes it!*

If this model makes sense, on every kilogram of meteorite, as an average, one of the ten spores that entered the atmosphere arrives still living on the surface of the Earth, that is to say, one out of the 10^{11} that left Martian soil.

For the time being, however, this is pure speculation. Even though in meteorites, organic material, even very complex one, is to be found all the time, nothing alive has yet been found in a rock fallen from the sky. Nor has anything alive yet been found on other planets, not even on Mars. But it is important to have demonstrated that, by making use of a realistic scenario, inside the solar system (or at least in its parts closest to Earth) and, starting from known forms of life, the mechanism of interplanetary transfer of life would not be impossible.

As a statement of principle, the same mechanism could be generalized, applying it to celestial extrasolar bodies. But even if it is possible that outside objects may have entered the solar system (in the same way that inside objects have for sure left it), we cannot, for the time being, assess, not even approximately, the frequency of their arrival.

A Very Special (Martian) Rock

On 7 August 1996, a press conference was held at the White House. President Bill Clinton opened thus: "Today this piece of rock", he said pointing at a rock shaped like a potato, "speaks to us and it does so through distances of millions of miles and times of millions of years: it speaks to us of the possibility of peace". Smiling next to him, in front of a crowd of excited journalists, there stood NASA director Dan Golding, with a rich group of scientists.

The bit of rock was a Martian meteorite. It had been the first one found in 1984 in the region of the Allen Hills in Antarctica: hence its name, ALH84001. The rock had arrived from Mars by itself, without any intervention on the part of the American space agency. But NASA had appropriated it because some experts would maintain they had found in it evidence of fossil life.

Probably it wasn't true, or rather the proofs were not sufficient, but it was likely. And the mere fact of saying so brought astrobiology to the attention of the great public, under the spotlight of media and generated for NASA the effect which Golding had been hoping for. Not by chance, the announcement was taking place exactly in time to allow the US Congress to touch up the financing of the budget of the aerospace administration.

But which were the elements of evidence? Figure 4.3 shows a detail of the photograph which at that time went all the way around the world. The structure at the centre, similar to a worm, had been interpreted as a somehow "fossilized" Martian form of life. A dispute got started, which is still carried on today, among experts of various disciplines to decide whether the structure be biogenic, the result of activity bound to elementary forms of life, or if it was simply of mineral origin.

Let us, first of all, make it clear that certainly ALH84001 has arrived from Mars. The small quantities of gas imprisoned in its pores have a composition identical to the one, well known, of Martial atmosphere. Moreover, the geological and mineralogical analysis leaves no doubts on the fact that the rock formed 4.5 billion years ago, exactly when Mars was forming. Much later, it surfaced from the inside of the planet to the surface, where it was for a long time in

Fig. 4.3 The side of the ALH84001 meteorite, which was cut in order to study its inner composition. The detail on the *right*, taken by electronic scansion microscope, shows a peculiar structure that could be a Martian microfossil

contact with liquid water; another proof of the fact that Mars, once upon a time, was not the desert it appears to be today.

Still much later, some 15 million years ago, as the consequence of the impact of a foreign body, the rock flew away escaping Martian gravity. It then took the rock quite a long time to arrive to Earth, but it made it: some 13,000 years ago it landed on Allen Hills.

Antarctica is the right place where to look for meteorites: not only do they well stand out, black on a sea of ice, but they also get preserved on a very little polluted environment of the terrestrial biosphere, where it rains very little and the temperature is constant. It is as if ALH had been kept in a freezer for 13,000 years, awaiting to be discovered.

Once it had been discovered and picked up with all precaution, AHL was immediately identified as coming from Mars, therefore of special interest. Apart from studying its geo-mineralogical history, the meteorite's organic compounds were looked for, actually finding many traces of carbonic material. It was nevertheless necessary to eliminate the suspect of pollution even in very clean Antarctica.

Analyses have shown that the content of carbon increases from the surface to the centre of the meteorite: in the case of terrestrial pollution, one would expect the opposite to take place.

Is everything clear, then? Unfortunately not. A lot of the organic material present in the rock is undoubtedly of terrestrial origin because it contains a sort of radioactive carbon which is downgraded and practically disappears in a few tens of thousands of years. Pollution, after all, has taken place.

Are we then back to the beginning? Not really: some organic compounds appear to be older and thus compatible with an extraterrestrial origin, but the puzzle is complicated and far from being solved.

But let us come to the "fossil wormlet". Its elongated structure as seen through an electronic microscope is very unusual; it had never been observed before in a meteorite. It looks like a row of contiguous segments, less than 1 μm long. It is remarkably reminiscent of certain terrestrial bacteria, which yet are usually at least ten times bigger. Recently, however, "nanobacteria" of the same size were found in the terrestrial underground, very deep down, alive and kicking.

There are also traces of an analogy with bacteria called magnetotactical that are sensitive to the Earth's magnetic field because they contain tiny crystals of magnetite. In the Martian sample too, crystals of this sort are present, and their geometry appears to be similar to the one of the crystals of a particular terrestrial bacterium, called M V-1.

The dispute between those who believe and those who do not, in the biological origin on ALH84001, is still on nowadays, and it has almost become an act of faith. Despite the existence of strong arguments in favour of the biological origin, what Carl Sagan, possibly the greatest planetologist of all times, said still holds true: *extraordinary claims require extraordinary proofs*. And the "extraordinary proofs in favour" are still to be found.

The solution of the enigma may come from the analysis of other Martian meteorites. Altogether, we have today 55 sure catalogued items, corresponding to a mass of 92 kg. There is thus material on which to work, although one could look back in regret on all those tons of rocks arrived from Mars and lost on the bottom of the oceans or maybe landed in a paddy field in Lomellina.

Encouraging results have been obtained by studying meteorite Yamato 593 (found on Mount Yamato, also in Antarctica), which shows many biomorphic structures similar to terrestrial bacteria. The hunt then is on and must go on, but the solution may be near.

Who knows who will be first: Martian life that lets itself be caught on Earth or terrestrial instruments that go and get it on Mars?

Chapter 5
Contact Astronomy: Ourselves, the Invaders

In the past 50 years, a new type of 'contact astronomy' was begun. We no longer sit and wait for bits of heaven to fall on our heads: our approach is now more active: we go to visit the celestial bodies from close-up and we descend on their surfaces and in some cases even bring back bits of them to Earth. It is a turnabout for humankind, made possible by the conquest of the access to space. Let's see how things went.

First Steps Outside Earth

The first physical contact between a man-made object and a celestial body dates back to 14 September 1959, when the Soviet probe 'Luna 2' touched the lunar surface. A few months earlier, the twin probe Luna 1 had missed the target by a few thousand kilometres, to become 'the first of the Sun's artificial satellites', as propaganda would state right away.

Luna 2 crashed when falling out of control at high speed but not by mistake: a 'soft' controlled mooning was still far away from the skills of Soviet astronautics. What makes it difficult is the fact that the Moon has no atmosphere, so parachutes cannot be employed: to slow down the descent, it is necessary to bring over from Earth braking back rockets and fuel and to have on board sophisticated electronic appliances to make sure the engines will be started at the right time.

Luna 2 was certainly a feat of propaganda, but it had the scientific and explorative value of showing that the human race was capable of

G.F. Bignami, *We are the Martians*,
DOI 10.1007/978-88-470-2466-3_5, © Springer-Verlag Italia 2012

hitting another celestial body 'on flight'. Half a century on, we tend to forget it, but it was a grand enterprise. Maybe 1 day, we'll go and visit the crater where it fell on the Moon, and among the deformed fragments of the probe, we'll find that red flag that Soviet propaganda declared to be on board, signed by Sergei Karaliov, the genius of astronautics. As a souvenir, it would really be special.

A few days after the impact of Luna 2, on 4 October 1959, exactly 2 years after Sputnik, that had begun the space age, the Luna 3 probe was launched, which 2 days later managed to enter an orbit around our natural satellite. Thanks to the images it sent down to Earth, humankind could for the first time see the other side of the Moon, which always turns towards us the same hemisphere. This too is an important time in the history of civilization; exactly on 4 October 1959 is to be dated the beginning of planetary exploration, that is to say, of our will and ability to visit the bodies of the solar system, first of all in order to understand them and then to explore them.

The first soft landing on the Moon was again achieved by the Soviets, in February 1966, by means of a probe that sent back to Earth the first beautiful images of the Moon's surface. In the scramble for the Moon, however, just at the crucial point, the Soviets were famously beaten by NASA's Apollo Project. Between 1969 and 1972, 12 US astronauts landed on the Moon and walked about on the Moon's surface (and also drove around on a vehicle, Fig. 5.1). They brought back home 380 kg of lunar rocks, collected here and there, in truth to tell; astronauts were then usually members of the military, not scientists.

Thus, the Soviet Union was devastatingly beaten in the race for the Moon. Later on, however, it succeeded, through miracles of robotics, in the not simple feat of collecting by means of automatic probes a few hectogram of lunar soil and to bring it back to Earth.

The lunar phase of the Apollo Project would appear as the beginning of a new era, but everything ended there. From 1972 to date, no human being has forsaken the Earth's gravity. Hundreds of astronauts of many nationalities have been orbiting (and some right now are in the International Space Station), but we are dealing with a low circumterrestrial orbit.

What has been happening to human exploration of space? Can such a long intermission—more than half a century will have elapsed

Fig. 5.1 December 1972: the geologist Harrison Schmitt, the only scientist among astronauts landed on the Moon, is pictured in front of a big boulder during the last lunar mission, Apollo 17. In the foreground, you can see the front part of the rover, the electric vehicle used by astronauts to move on the surface of the Moon

between Apollo and the next human extraterrestrial adventure, if ever there is one—mean that we have turned back, that we are no longer able to think big?

Maybe, but more probably, the case of the Apollo Project was exceptional, born of a combination of political factors (the Cold War), human factors (the union of John Kennedy and Werner von Braun), emotional factors (the promise made to a murdered President) and cultural factors (the drive of country and new NASA).

At the time, risks were taken that today would certainly be unacceptable and sums were allotted that could not be suggested nowadays. Maybe Apollo was the exception and the present rhythm of space exploration is the rule, although it seems to be advancing very slowly if compared with the glorious Sixties.

Going to Venus and Mars

Even during the lunar programme, the two big space powers, USSR and US, were beginning, albeit among great difficulties, planetary exploration beyond the Moon by means of automatic probes aimed at studying planets.

The Soviets, well aware of the propaganda worth of space, having at their disposal powerful vector rockets in practically unlimited amounts, tried right away to hit big towards Mars (minimum distance from the Earth 55 million kilometres). But in vain between 1960 and 1962, they launched at least five probes, all of them with no success.

In those same years, the USA, which at the beginning may have had more limited means, launched the Mariner series made up of the first probes designed to reach other planets. The first Mariner failed, but the second, somewhat to everybody's surprise, worked: it arrived at just a few tens of thousands of kilometres from Venus, sent down to Earth just a few data, but it was a big psychological success.

Then, it was Mars's turn: Mariner 3 failed, but Mariner 4 in 1965 arrived a few thousand kilometres from the red planet and from there sent 22 images: they were the first photographs of a planet as seen from close-up, and they had a great impact on the whole world. These first NASA images of Mars carry an historical significance: they refuted once and for all the legend of the supposed Martian 'canals' born a few decades earlier, as we shall see further on, out of a curious misunderstanding.

In 1967, despite the premature death of Karaliov, weakened by years of gulag, which he had been compelled to undergo by Stalin's purges, the Soviets achieved another spectacular *first*, reaching Venus (minimum distance from the Earth 40 million kilometres): the spacecraft Venera 4 managed to parachute a probe into the thick Venusian atmosphere. Then, in 1970, Venera 7 managed to come to land on the Venusian soil: it was the first human artefact to touch the surface of another planet.

The data sent back to Earth have caused us to discover that Venus is a hellish place, with a pressure equal to 90 terrestrial atmospheres and a temperature at ground level of 450°C: hard to imagine one could live there.

Fig. 5.2 The surface of Venus seen in 1982 by the soviet spacecraft Venera 13

The photographs taken by the probes of the Venera Programme (Fig. 5.2) remain to date the only images we have of the surface of Venus.

In the 1970s, the race to Mars was taken up again. In 1971, the USA put around the red planet Mariner 9, which became the first artificial satellite of another planet and sent down to Earth lots of beautiful images. The Soviets would not give up: on 2 December 1971 touched the Martian soil and from there sent a few data, although only for a handful of seconds; in any case, it remains the first man-made object to land softly on the surface of Mars.

The first physical contacts with three bodies closest to us in the solar system (Moon, Venus, Mars) have all been the achievements of Soviet space research. With these historical events there closed the first phase of exploration 'by contact' of the planets around us, but not only: an historic-political era of our civilization came to an end. In the final 30 years of the twentieth century, the technological gap in robotics and electronics between USA and USSR was bound to keep widening, all the way to the fall of the Soviet Union itself: it could even have been one of its important concurring causes.

After the Seventies, Russians have concentrated their space efforts on missiles and on human flight, areas in which they are still today the protagonists, 20 years after the fall of the Berlin Wall. Aware of not standing a chance in a competition based on more and more refined and far-fetched technologies, they know they can manage much better with cosmonautics: after all, electronic circuits can be miniaturized, astronauts cannot.

Our Print on Other Planets

When people go on a picnic, it is a good rule to bring back leftover food and rubbish. But for sure in the human exploration of the solar system, we were not so well behaved. Here are some statistical data.

On the Moon, in half a century of 'terrestrial bombardment', we either placed down or let fall some 70 big objects, to which are to be added countless bits and pieces (from screws and bolts to photos of astronauts' children). Altogether some 180 tons of terrestrial iron-work and electronics placed down on the Moon by USSR, the USA, Japan, Europe, China and India, as compared with 380 kg of lunar rocks picked up and brought down on Earth by the Apollo Project (plus some 2 kg of the Luna Project, the Russian robotic mission). The biggest objects left on the Moon's surface—what was left over from lunar modules Apollo and *Lunakod*, the automatic Russian rovers—we can today take pictures of from the lunar orbit thanks to NASA's Lunar Reconnaissance Orbiter.

On Venus, from 1969 to date, there have either landed or hovered in its thick atmosphere to fall eventually to the ground, some 20 human objects of a remarkable size (mainly Soviet ones, but also a pair of giant French stratospheric balloons) for a total mass of above 22 tons.

On Mars too from 1971 to date, we have put down a lot of stuff: some ten major objects, chiefly American, plus a 'missing' ESA probe, which may have fallen on the planet, total mass of our handcrafts on the surface: some 8.5 tons.

Here and there in the solar system, we also 'touched' one comet and two asteroids. In 2001, NASA's NEAR probe landed on Eros asteroid: 487 kg left on that small world. In 2005, poor Tempel comet was bombarded by a shot weighing 370 kg sent off at high speed by Deep Impact probe. Again in 2005, the Japanese mission Hayabusa dropped on the asteroid Itokawa a tiny lander—only 591 g—but a very ambitious one: it has collected (perhaps! analyses are still being carried on) a pinch of asteroid dust, which the probe then took back here landing in Australia in June 2010, after a fantastic voyage of six billion kilometres there and back.

Moving to the outer planets, the massive probe Galileo (2,564 kg) has been made to sink into Jupiter's atmosphere in 2003, after a glorious mission to explore Jupiter's system which lasted 8 years. On the surface of Titan, the biggest of Saturn's moons, in 2005 ESA's probe Huygens landed (350 kg): it is the contact with the celestial body furthest away ever reached by humankind; we can be proud of the fact that it was a European achievement.

Thus, altogether contact astronomy has taken all of 213 tons of man-made material from Earth here and there into the solar system, on the surface of seven different objects, not bad for an invasion, in something like half a century.

One should add that it is not probable that all such material was ever accurately sterilized before departure, especially in the case of early lunar missions and of the interplanetary probes of the Seventies. Who knows how many terrestrial microorganisms we have been spreading around the solar system.

We are thus competing with the natural mechanism of exchange of interplanetary material; instead of kinetic energy supplied by cosmic impacts, we employ the chemical energy of our vector rockets.

More Close Encounters

Apart from the cases in which there has been a physical encounter, very numerous have been the visits in the vicinity of other planets: the four inner rockies Mercury, Venus, Earth and Mars with their dearth of moons (only three for four planets) and the four gas giants, with their crazy abundance of moons, which it is probably impossible to count (to date we are up to 63 for Jupiter, 62 for Saturn, 27 for Uranus and 13 for Neptune).

For planets and a large part of their satellites, we have today a data archive made up of millions of photos and surveys performed by orbiting instruments. And we have also collected close-up images (from a few kilometres to a few thousands of kilometres) of several asteroids and of at least four comets. It is an impressive archive, difficult to be summarized and in which one must learn to find one's way. One can, among other things, draw information there for the study of prebiotic materials and conditions or somehow connected with the development of life.

Further on, we shall discuss in detail both Mars and the comets, but interesting surprises have also come from the exploration of Saturn's satellites. Titan, bigger than the planet Mercury, is the only moon in the solar system that has a thick atmosphere. The Cassini–Huygens mission—born of a collaboration between NASA and ESA—has analysed it, finding it denser than the Earth's and rich in organic molecules. Huygens also took photographs of the surface where liquid rivers flow and lakes form (Fig. 5.3, on the left). It is very cold on Titan, but nevertheless, we found on it a real incubator of organic material.

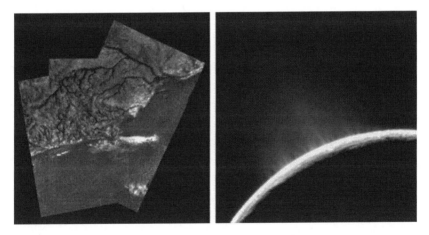

Fig. 5.3 On the *left*, a mosaic of images taken by Huygens probe during the descent towards the surface of Titan; a 'river basin' can be descried, tributary to a great lake of liquid hydrocarbon. On the *right*, gas jets issuing from Encelado, another of Saturn's satellites

Encelado too, another albeit much smaller frozen moon of Saturn, seems to hide inside itself something promising. Cassini has observed gas jets that come out of its Southern polar zone (Fig. 5.3, on the right), where the temperature seems higher than usual and where there could even be liquid water. For sure, water is contained in the vapour jets emitted by Encelado, where simple organic molecules, such as methane, are also to be observed. Here, as on Titan, environmental conditions could be compatible with the development of forms of life.

Very interesting have also appeared the data coming from the Galileo probe regarding Jupiter's 'Medici' satellites (the four bigger ones, discovered in 1610 by Galileo, who dedicated them to Lorenzo de' Medici). In particular, the frozen surface of Europa shows cracks that make us think of a moving, and therefore liquid, underground ocean. Who knows that inside it something may swim.

Something of the sort can be said of Ganimede, which with its 5,200 km diameter is the biggest moon in the solar system, even complete with a small magnetic field of its own. Ganimede has a mass compatible with the existence of a still warm core, which could originate a volcanic activity. Thus, although covered by a crust of ice 100 km in thickness, below that there could be liquid water. If there were also underwater volcanic openings to supply energy and nourishment, the situation would be similar to be found on the bottom of our oceans: around them forms of life thrive very different from the ones on the surface, with organisms that manage very well without neither light nor oxygen.

But the active exploration of the solar system has been going—and still is—also beyond Jupiter and Saturn. Four NASA missions (Pioneer 10 and 11 probes, launched in 1972–1973, and Voyager 1 and 2, launched in 1977) are even by now out of the solar system. Both Pioneers did not carry on board filming appliances, but the two Voyagers did and to them we owe our images of the two outer gas giants, Uranus and Neptune, with their systems of satellites.

The only thing missing is a visit to Pluto. For the past few years, it has no longer been a proper planet, but this does not diminish its scientific interest; it could, for instance, be covered in a thick hydrocarbon snow. NASA's probe New Horizons, which left Earth in 2006, will reach Pluto in 2015 to carry on and leave the solar system for good in 2029.

Thus, with New Horizons, after the Pioneers and Voyagers, there will be five pieces of human handwork which left our planetary system and are travelling starwards. They will keep getting further and further away, until either some gravitational interaction or an improbable chance bump may waylay them. They will carry on with their motion even when the human race as we know it will be extinct on this planet. We have been here two million years: in two million more, shall we still be around?

In any case in five billion years, the Sun, in its inevitable transformation into a red giant star, will be blown up until it swallows Earth, which will be vaporized into nothing.

At that point, it will only be those flying objects, by then arrived who knows where, that bear witness to the ancient existence (of old) of an intelligent species by now extinct.

Is There Life on Mars?

When life is sought on another world, it is difficult not to be anthropocentric or at least geocentric, though knowing it should not be so.

We can imagine forms of life completely different from terrestrial ones, for instance, with silicon instead of carbon, with ammonia (or methane?) instead of water, with unexpected chemical reactions and so on using our imagination. At the point of choosing a strategy to look for evidence of life on another planet, however, we fall back again and again on what we know for sure: that is, that life, a life like ours, requires carbon, water, solar, or at least thermal, energy and nutrients in the soil.

Of all these ingredients, water is certainly the most obvious and the easiest to look for. Thus, even centuries ago, when with the first telescopes men would observe spots on the Moon and on Mars, these would almost automatically be called 'seas'. In the case of Mars, the hope of finding liquid oceans on the surface rapidly diminished with the first photographs from the Mariner probes at the beginning of the space age, vanishing all the way with the arrival of the first probes on the planet's surface.

As for the exploration of Mars, a bitter truth should be voiced, 40 years on from the first landing there (after dozens of missions, among which many failed ones). We have collected and comfortably analysed at home more than 50 pieces of Mars arrived by themselves on Earth; in the meantime, we have also deposited, at a very high price, some 10 tons of material on the planet, running the risk of polluting it forever. Well, after all this, we have not yet found life on Mars, maybe because there isn't one. What is worse is that we cannot either rule out the presence of biologic material, either living or dead, on the surface of Mars.

The first researches into life on the surface of the red planet were performed by the twin Viking probes, landed on different spots of the planet in 1977. The Vikings carried on board all of four devices inspired by the great Carl Sagan and dedicated to the search for biological traces, also inert ones, in the soil. The instruments, which would work perfectly, were of course looking for forms of life based on carbon and water.

The idea was to collect a small quantity of the sandy soil in front of the probe (which was not self-moving) and bring it inside the probe for an accurate chemical analysis, based, for instance, on the observation of possible reactions when the sample would be placed in a humid environment or in one with nutrients (brought over from Earth).

None of the four experiments produced a positive result. The matter has long been discussed, and someone still maintains that one of the instruments on board the Vikings has in actual fact found something biological in the collected material. In every instance, it is the case of very indirect and garbled indications, quite far from supplying those *extraordinary proofs* that are necessary to be able to assert for sure that life on Mars has been found.

It should be said that only a few spoonfuls of sand were examined, collected in a couple of spots haphazardly picked. If one repeats the same experiment on Earth in a desert at high altitude, exposed to ultraviolet radiation, similar results are obtained.

Carl Sagan himself, the mind behind the Vikings, had no clear ideas on what to look for. At a certain time seeing that TV cameras, just landed, would not show lizards and chicken moving about in the Martian desert, he jokingly suggested that the following mission should be equipped with spot lights to be turned on at night: the Martian fauna is nocturnal and hides during the day (Fig. 5.4).

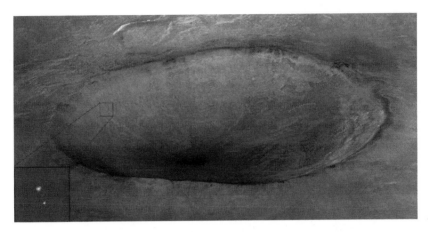

Fig. 5.4 15 May 2008: the descent of Phoenix probe, with Heimdall crater in the background, not far from Mars's North Pole; this was taken by the videocam of Mars Reconnaissance Orbiter, orbiting around the planet

The fact is that since the Vikings, 30 years have elapsed before another active probe could be had on the surface of Mars, one able to search in earnest after biological traces. We are talking of Phoenix, a well-designed mission, made to land in the polar zones of the planet, where certainly there's water in the guise of ice, but with temperatures that in the Martian summer rise above zero.

The mission lasted few months, starting in March 2008 till November (during the terrible polar winter, similarly to Earth; on Mars too, the Sun goes down and the source of energy is thus missing).

Phoenix was equipped with instruments of the Vikings' sort, not much more. Its automatic shovel (Fig. 5.5) could dig the soil only a few tens of centimetres, so it only collected chiefly superficial samples.

Phoenix found nothing biologically interesting. It rather identified an inorganic compound which had gone unrecognized 30 years earlier and which might explain in completely non-biological terms some still doubtful results from Viking.

Between the time of the Vikings and the Phoenix mission, the exploration of Mars has had mixed results. From 1975 to 1981, in the gap between the Apollo Project and the first Shuttle flight, NASA was mainly engaged in rebuilding the possibility of human access to space in view of the construction of the Space Station. Then, there are

Fig. 5.5 A solar panel of the Phoenix lander and the robot arm, which, having picked a sample of Martian soil, takes it to the analysing instruments on board the probe

a series of failures, even embarrassing ones, such as when a probe aimed at Mars went astray because software on board had got feet mixed up with metres.

Over some years, the most important result in the search for life on Mars was the indirect one, obtained in 1997 by the orbital probe Mars Global Surveyor: the first mapping of hydrogen on the surface of the planet showed as believable a diffuse presence of water on Mars. The datum was later confirmed by Mars Odyssey, orbiting around Mars since 2001.

After 1997, surface exploration started again, first by the Pathfinder probe and its tiny mobile robot Sojourner, then in 2004 with the self-moving robots Spirit and Opportunity, which were much bigger. Regrettably, these do not carry on board instruments specifically aimed at the search for life and to date have not offered particularly interesting results on this point. They have, however, further confirmed in situ the evidence of a past with liquid water on the surface of Mars, confirmed by the presence of sedimentary rocks like the ones in Fig. 5.6.

Fig. 5.6 In this photograph, taken in 2006 by the Opportunity rover near the Herebus crater, one can recognize Martian rocks with clear stratification of the sedimentary type

Another important hint of the possibility of Martian life—a life based on carbon, be it clear—came from an European mission, Mars Express, orbiting around Mars since 2003. An instrument on board this spacecraft, whose responsible person is an Italian, Vittorio Formisano, discovered localized evidence of methane in the atmosphere of Mars. This is a very relevant datum: on Earth, the great majority of methane to be found in the atmosphere originates from the metabolism of bacteria, for this reason called *methanogen*; a remarkable presence of methane on Mars would make us think of a similar origin. Moreover, in Mars's atmosphere, methane cannot survive intact longer than around 400 years; if we find it today, thus, this means that a source of some type exists that is producing it all the time.

The question then becomes as follows: what is it that produces the methane to be found on Mars? Because the presence of methane is sure, the Italian observation by Mars Express has been confirmed by independent measurements made by means of a ground telescope.

It is difficult to measure precisely the mass of gas present in specific zones of Mars's atmosphere. Recently, in the region called Nili Fossae, localized methane was found, localized at a concentration of 60 parts per billion. It may seem little, but reckoning a bit, it means that a mass of some tons of methane gets pumped every year into Mars's atmosphere. On Earth, a few thousand cows would suffice to generate it: it isn't such a big production, but neither is it a negligible one; it is difficult to explain it without attributing it to the metabolism of living beings.

The biological explanation will become almost mandatory if another observation is confirmed—a bit on the limit of instrumental sensitivity, truth to tell. This was made by Mars Express in the atmosphere of Mars, and it entails the presence of formaldehyde. In itself, this is no surprise, since it is a natural product of methane's oxidation. But formaldehyde is much more unstable than methane: in Mars's atmosphere, it can only survive a few hours before downgrading. To conclude, if there is any formaldehyde, this means that the quantity of methane which is present at every time on Mars must be much bigger than up to now estimated, and it seems impossible to justify it if not by a microbic origin.

To summarize, after 40 years on Mars, we have a few hints, but nothing certain yet, on what regards the presence of life or at least a life similar to the one on Earth. We can yet trust the fact that the coming decade can bring great progress, thanks to two missions planned on Mars's surface. In 2012, it will be NASA's Mars Exploration Rover, a very advanced robot, complete with instruments for analysing both the atmosphere and superficial rocks. Then, by 2018, ESA will arrive with Exomars: it will be the first mission which will be able to drill Mars's surface down a couple of metres, thus implementing a tridimensional sampling of the biosphere. Exomars will explore environments that are screened by the lethal effect of radiations and where water could be rather plenty, possibly in liquid form, who knows... allow us to dream.

Chapter 6
Contact Astronomy: Comet Dust

In our solar system, comets are billions, possibly hundreds of billions. They make up a huge, unknown population of primaeval objects, which is collected in an outer zone which may extend up to a light year from the Sun.

The objects we call comets are made up of a mix of rocks, ice and volatile materials: they are the same elements that made up the base for the formation of planets.

From time to time, chaotic gravitational encounters push one of these objects into the inner parts of the solar system, placing it in a way of falling onto the Sun. The Sun's heating sublimates the outer strata of ice, thus generating around their nucleus a sort of rarified atmosphere called 'tail' (the more appropriate name for which should be 'hair', from the Latin 'coma', hence 'comet'). The pressure due to radiation and to the solar wind also shapes a 'tail' which can be tens of millions of kilometres long and which points in a direction opposite to the Sun.

Many comets fall into the Sun and forever disappear. Others enter an orbit around our star and later on can thus repeatedly pass also close to the Earth. If we are lucky and this happens during our lifetime, for many nights we'll be able to see with the naked eye one of the most beautiful sights in nature.

G.F. Bignami, *We are the Martians*,
DOI 10.1007/978-88-470-2466-3_6, © Springer-Verlag Italia 2012

From Giotto Comet to the Bombarded Comet

In 1301, Giotto di Bondone, with the eye of a revolutionary realist painter, observed exactly the spectacular passing of a comet. It was the one we today call 'Halley', after the name of the English astronomer Edward Halley, a pupil of Newton's, who in 1705 foretold the periodic returns of that comet. Giotto painted it in the 'Cappella degli Scrovegni' in Padua, a flaming light just as he had seen it. It was a new direction in painting and, more generally, in European culture; for this reason, ESA gave the name of Giotto to the probe that went to meet Halley's comet during its passage near the Earth in 1986.

The mission was a great success for European space industry, which succeeded in beating Soviet and Japanese competition, while NASA for once remained at the starting blocks. But also from a scientific point of view, it was a real scoop: Giotto sent to Earth more than 200 images taken during its approximating, up to less than 600 km from Halley's nucleus. We thus discovered that inside the sparkling tail, there is a nucleus black as black can be, roughly as big as the island of Ischia, with emission jets localized on the side exposed to the Sun.

The colour black was a spectacular confirmation of what had been foretold by Fred Hoyle, who imagined comets like dirty snowballs (or rocks covered in snow, depending on the point of view), covered by a veneer of organic substance similar to tar, a slightly simplistic model maybe, but in fact one compatible with what has been observed. The first remarkable closeup images of a comet's nucleus confirmed that it contains abundant organic material.

It was then necessary to wait for more than a decade in order to have a chance of a new space contact with comets, but this time, NASA was there in all its splendour.

The Deep Space 1 probe had left in 1998 as a 'technological mission', that is one aimed at testing new types of driving and instruments; as sometimes will happen, it was a solution in search of a problem. The mission extended to 2001 and till then scientifically not very significant was nonetheless bound to a brilliant *finale*. It was decided to try for a close encounter with comet Borrelly, an object 8 km long and 4 km wide, very similar to Halley: the outcome were high-resolution images even more finer than Giotto's, although taken

Fig. 6.1 Tempel comet photographed by Deep Impact probe, after the impact of the 350-kg cartridge shot by the same probe

from 2,000 km away. The comet's nucleus, more quiescent than Halley's and thus more visible, showed the by-then expected variety of colours and shapes, thus confirming the nature of comets' surface already observed by Giotto.

The most spectacular contact, however—in some people's opinion even outsized—between a human object and a comet took place on 4 July (not a chance date) 2005. NASA's Deep Impact mission, which had started 1 year earlier, once arrived in the neighbourhood of the comet, sent off an 'impactor' made of some 350 kg of brass, which also carried on board a driving system and a camera. After a perfect approach, on Independence Day, the suicide probe crashed at high speed on the comet's surface, after sending to Earth a final image taken 3 s before impacting.

The impact made a crater, on the comet's surface, as large as a football pitch, raising a cloud of gas and dust (Fig. 6.1), which, lit by the Sun, could be analysed by the mother probe, but also by other telescopes, both in space and on the ground. Here too vast evidence was found of simple organic carbon and nitrogen compounds, apart of course from abundance of water. No trace of big complex organic molecules, but these are fragile and they might have been destroyed in the impact.

A Bit of Hair Caught and Brought Back to Earth

To date, the most important contact with a comet has been the fourth in the history of humankind, realized by NASA's Stardust mission. After Apollo's (and the Soviet programme Luna's) rocks, by Stardust, extraterrestrial material was brought on Earth for the first time since the 1970s, by means of a purpose-planned space mission. In this case, it concerned traces of comet dust.

Stardust left in 1999 and, after a long interplanetary cruise in 2004, reached comet Wild 2. The mission had been planned to catch and bring home a pinch of comet material, being careful in obtaining it intact. It is a difficult achievement, especially because comet dust moves at high speed (even tens of kilometres/second) if compared to the probe and it thus risks 'burning' in the high temperatures generated by the normal impact: it would thus make it to search for the most complex and delicate molecules.

For this reason, Stardust had on board a battery of trays containing a special *aerogel* that would capture the particles of comet dust. It is an extremely porous gel based on silicon, solid but almost as light as air: it's very low density allows the gel to absorb quick particles in a soft and gradual way, without too much heat being dissipated (Fig. 6.2).

Once the passage by the comet has been accomplished and the special containers of tail have been put outside, with a manoeuvre almost reminiscent of science fiction, Stardust turned about and 2 years later, in 2006, it was again close to Earth. Here an appropriate capsule detached itself from the probe: it was perfectly sealed and thermally insulated, and it came back into the Earth's atmosphere, and, by means of a parachute, it safely landed on a desert in the United States.

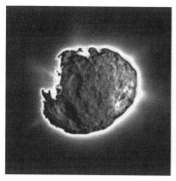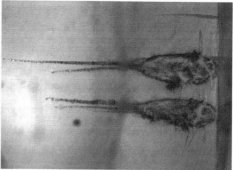

Fig. 6.2 On the *left*, the nucleus (core) of Wild 2 comet photographed by Stardust probe. On the *right*, a microscope image of the traces released in Stardust aerogel by two particles of the Wild 2 tail; indicators point to the grains of 'comet dust' brought back on Earth by the probe

It took then 3 years to analyse all the way through the composition of the comet dust (and not only the one from the comet) captured by aerogel. With a gesture of great generosity and scientific value, NASA decided to share with researchers the world over the dust thus collected, handing around samples to over 150 research labs.

The most important discovery was announced in August 2009: in the dust, glycine was found, which is one of 20 amino acids that, combined in the most peculiar ways, make up the millions of different proteins, out of which are living beings on Earth made. And we can be sure that it is an extraterrestrial amino acid, because the isotopic composition of carbon taken home by Stardust is different from the one which is typical of our planet.

The Stardust enterprise somehow closes a circle which was opened in the 1970s with the finding of the Murchison meteorite, rich in amino acids arrived onto Earth by themselves. In the case of the comet too, we are faced with advanced and complex organic material, but a 'pristine' one, which was shaped before Earth, or in any case independently of it, maybe even outside the solar system.

The epoch-making result obtained by NASA with Stardust will be difficult to beat, but we wait confidently for the outcome of the European mission Rosetta. It started in 2004. In 2014, it will spend months orbiting around a comet in the outermost zones of the solar system. If everything goes well, it will also send down a small robot probe called Philae, which can analyse in situ the surface of the comet down to a depth of at least 1 m, thanks to a sample drill manufactured in Italy.

It will be the first 'live' taste of a comet. Rosetta will send us live (with a slight delay due to the transit time of signals that 'only' travel at the speed of light) the taste, the texture and the composition of a comet met when it is still faraway from the Sun, that is, in the same condition in which it has been for the past four and a half billion years. Wow.

Sun Dust on Earth

After studying the composition of the bodies in the solar system which we visited, including comets' tails and meteorites delivered to our doorstep, what remains to be understood is the relation between the large amounts of the various elements measured in the rocky bodies of the solar system (Earth included) and that of the Sun itself.

Since we cannot think of sending a probe to fetch a few spoonfuls of solar plasma back home, NASA has entrusted the mission Genesis with the task of collecting a bit of solar wind, the flux of particles that is emitted by the Sun and spreads in interplanetary space. These subatomic and atomic particles get emitted at high speed (hundreds of kilometres/second) by the solar corona of which they somehow represent an extension that goes so far as to contain both Earth and the other planets. The aim is exactly that of capturing Sun dust.

Launched in August 2001, Genesis in 3 months reached its goal, 1.5 million kilometres away: a very special position in the Earth–Sun gravitational system called *point of libration*, where the forces that act on the probe are balanced, allowing it to remain forever in line with Earth and Sun.

Here, so very far from our atmosphere and free from any terrestrial interference, Genesis deployed all its sensors, exposing them to the solar wind. They were trestles of panels of a special composition, worked out in a differentiated way for the various atomic species that were expected to be collected.

Genesis collected particles of solar wind uninterruptedly for 850 days, until 1 April. Then, the panels were safely closed into the hold of the capsule, which had been designed to take them to Earth, and the journey back started.

The sequence of the way back was rather spectacular. In order to bring down to a minimum the risk of terrestrial contamination of the most precious samples, the capsule, after being braked by a parachute, should have been caught in flight by a team that was expecting it on a helicopter. Unfortunately, the malfunctioning of an accelerometer did not give the signal of parachute opening, turning the controlled landing into a crash on the Utah desert.

Panic at NASA: what had happened to the samples of solar wind? Would it be possible to recuperate something? The fragments of the capsule—10,000 bits and pieces—were lovingly picked up and cleaned from the dust and from the salted sand of the terrestrial desert. Then, the analysis began of the particles brought home. With a big sigh of relief at that point, it became clear that the most important data were safe: at least some of the containers had remained airtight.

The first thing to be studied was the isotopic composition of the particles: as we already saw, what appears to differentiate chemical element on Earth from the extraterrestrial ones is the mix of isotopes that make them up, and it is just to understand the reason for the difference that it was decided to analyse the source of those elements, the Sun.

Genesis' surviving containers, as luck would have it, allowed us to measure the isotopic composition of the various elements contained in the solar wind. The clearest data with regard to oxygen and nitrogen is as follows: without any possibility of doubt, their isotopic composition in the solar wind is similar to the one found on meteorites, on the Moon, on Mars and on Jupiter, but it is different from what is measured on Earth. The same seems to be true for carbon and possibly also for other elements.

The result is striking: our planet would constitute an anomaly in the solar system. It appears improbable, but then we must understand what happened on Earth that did not happen elsewhere. Because in any case at the beginning, there are always the Sun, the proto-planetary disc and the mother cloud: from there, we too undoubtedly come.

A good number of theories have already been suggested on how and why the isotopic composition of many terrestrial elements (at least the light ones) has been modified regarding the wind of particles emitted by the Sun. The hypotheses aired are the most varied: from the influence of a supernova (at present excluded), to the erosion by the solar wind, all the way to the fall of comets on Earth.

Yet from the Sun, no molecules arrive, much less organic molecules: sure, we are the children of the stars but in a more indirect way; molecules, especially the more complex ones, do not come from our star but from further away. Who knows wherefrom.

In particular, a recent study on comets opens up new horizons. It is possible they were formed at the time of the birth of the brood of stars that included the Sun; and maybe at the beginning, they were 'exchanged' among the group of stars still close to each other. When later the siblings of the Sun went each its own way, each brought with itself a few comets.

The numberless comets to be found today in our solar system could thus not have been born together with our Sun but rather come mostly from other stars. In this case, they would be primaeval samples of a whole generation of stars that at birth could have characteristics different from one another.

Comets would then be even more interesting messengers than we thought.

Chapter 7
From Bricks to House: What Is Life?

Introduction

We have now reached our most difficult point. In notes that purport to cover all the ground from Big Bang to our own time, by way of Mars, a short summary of what we know about life is de rigueur.

The bricks of life have appeared to us in interstellar clouds, sown upon and enriched by stars, which, by means of their nucleosynthesis, give us the magic cocktail CHNOPS. We have followed those self-same bricks, that is to say, the big organic molecules of amino acids, up to their arrival into our solar system (and probably into other planetary systems as well). They are an integral component, albeit small, of the matter that, from the initial cloud, collapses to form the star and—something which is more significant for us—to also form the protoplanetary disc and then the planets themselves.

With comets, asteroids and meteorites of every sort, even Martian ones, we have also seen that for sure the bricks of life have been reaching Earth from the very moment it could be inhabited, that is to say, longer than four billion years ago.

Yet, up to this point, we have not witnessed anything 'alive' arriving on Earth from space, whether on board comets or meteorites. We see bricks, not houses, raining down from the sky: we see amino acids, but neither proteins nor, a fortiori, ready-formed cells in working order. Thus, we do not yet know for sure whether we are the Martians. Yet, that life on Earth exists is an undeniable fact.

The time then has come for us to attempt to understand what life is and how life began. The two enterprises—we should say it right

G.F. Bignami, *We are the Martians*,
DOI 10.1007/978-88-470-2466-3_7, © Springer-Verlag Italia 2012

away—belong to the hopeless sort. Hosts of famous biologists, a few Nobel Prize winners included, have had their go in vain. In the current opinion of experts, life is such a complex phenomenon that it cannot be encompassed by a definition, especially because such a definition should be at the same time simple, exhaustive and unambiguous. Understanding how life began is an almost impossible challenge, since its beginning has not left any trace.

From our point of view, it is at any rate important to summarize the few things we know on this topic (and the many we don't know) and try to imagine what can have happened from the moment bricks appeared on Earth to the coming together of the house.

During the past 50 years, biology has moved significantly forward in the direction of understanding the molecules of life. Today, we know that the information which serves to make up a living being is stored in DNA: this is done by means of proteins, that are, at the same time, the structure and the operational tools of any biological cell. RNA, in turn, is the conveyor belt between DNA and protein: it allows us to read, translate and regulate the coded messages coming from genetic material and to implement them by means of proteins.

The knowledge we have acquired on these molecules and on their operational mechanisms takes us well beyond plain space bricks. Yet this is not enough. Although we may take these well-known ingredients and put them together, we don't even go near making up the very simplest living being, such as a bacterium.

The universe, on the contrary, has been successful all the way through, scoring another important goal in the match which opened this book, a match which goes on today as lively as ever. The universe has clearly been able to make a human being, but human beings have not yet learned how to make themselves up from scratch in a laboratory.

Richard Feynman used to say: 'What I cannot create I do not understand'. Yet, the irrepressible drive of human beings to create has already made a few remarkable things happen. In a display window of the Royal Museum of Scotland, for instance, a stuffed sheep is on display, called Dolly. Born in 1996 at the Roslyn Institute in Edinburgh thanks to the efforts of Ian Wilmot and colleagues, Dolly was the first mammal created starting from an adult body cell. If we cannot yet create life from zero, we have at least started to reproduce it from known elements.

Almost 15 years later, a further step forward was taken by succeeding in making something that was not there. J. Craig Venter and his team, in Maryland and California, created a synthetic copy of the chromosome of a bacterium and moved this copy into a bacterium of another species, thus substituting its DNA. As a result, the cell with the synthetic genome began to reproduce itself and operate under the direction of the new exogenous genetic material.

If we think that this achievement is to be added to the reading of the complete human genome, accomplished in 2003 by the same research group, it is clear that the new feat by Venter, built step by step on the accumulated genomic and bio-computational knowledge, is a good illustration of Feynman's words.

After this extraordinary achievement, what results await us from what we now call synthetic biology? Could it possibly be synthetic vaccines against flu and infective diseases, especially those we cannot yet prevent, or maybe intelligent bacteria or algae that eat up oil or the CO_2 surplus in our atmosphere and excrete petrol? The possibilities that open up appear to be infinite. As the saying goes, at least among astronomers, 'sky is the limit'.

It would not be right, however, to say that Craig Venter created life from scratch: his synthesis of a new species began from the known elements of an existing species. For this reason, it would also be wrong to accuse him of 'playing God'. It's a simplistic phrase which has been used time and again over past centuries against all those who carried scientific progress beyond the limits of common knowledge. Venter's result takes us however a bit closer to understanding the origin of life, now that we can 'read it and write it' and that we are beginning to reproduce it, albeit in very simple forms.

What Is Life?

In February 1943, right in the middle of WW2, the Austrian physicist Erwin Schroedinger gave a series of lectures at Trinity College, Dublin, in non-fighting Ireland. The general title of the lectures, which were immensely successful, was 'What is life?' A few months later, in 1944, Schroedinger collected the text of his lectures in a book by the same title, which was immediately translated and published in many international editions and influenced a whole generation of young researchers.

Schroedinger had won the Nobel Prize for physics in 1933: why would then a scientist, already very famous, investigate such a topic as life? Possibly because such a clearly haunting topic was considered a real mystery by most people. Schroedinger was the first to suspect that the answer to some of the fundamental questions on the nature of life would soon come thanks to knowledge methods stemming from chemistry and physics. In the end, science would confront and possibly solve, by means of its modern tools, the so-called mystery of life.

As an example of the problems to be faced, Schroedinger, as the good Austrian he was, had chosen the 'Hapsburg lip', a protrusion in the lower lip that would be transmitted from one generation to the next in the Imperial family. How was it possible, Schroedinger wondered, that a similar character had been preserved across centuries? After all, the disorder introduced by thermal random motion, which all molecules undergo, biological ones included, should have done away with the transmission of this hereditary peculiarity. Schroedinger's example introduced, with great foresight, genetics' fundamental notion: genes were, in his view, the holders of a coded message.

We shall not move too far into Schroedinger's hypotheses. Suffice it to say that his ideas on the genetic code and on life as a mechanism for transmitting information, and thus also as an order against thermodynamic disorder, were soon to be experimentally demonstrated. Only 10 years later, in 1953, James Watson and Francis Crick would show the world the elegant double-helix structure of DNA. After another decade, in 1966, the genetic full code would also be revealed by, among others, Marshall Nirenberg, Sidney Brenner and again Crick.

More than half a century from these fundamental discoveries, however, we are not yet able to define the universal characters of living beings. For this reason, it still is impossible to really create life from scratch or at least from a small glassful of CHNOPS, which ought to be readily available and not even be too expensive.

We could maybe console ourselves from our inability to define life with a few theoretical remarks. Living beings are *complex systems*, that is to say, systems that show to be internally organized in a series of *emerging properties*, that is to say, properties that only show up, emerge, from the whole, without the possibility for any one of the parts, taken by itself, to individually show such properties. Examples of these properties are the fact that living beings are self-sustaining, can reproduce themselves, can grow, organize themselves in orderly cellular structures, interact with the environment outside them, evolve by means of mutations, hold and transmit pieces of information inside biopolymers and have a metabolism. A particularly interesting characteristic of living beings is that their biological molecules often are *omochiral*, just like the molecular bricks we've seen coming from space, while the same 'mirror' molecules—their *enantomeres*—either are not to be found in nature or are very rare.

The existence of emerging properties that we cannot yet describe is also possible. Here too, there's an 'official' justification, taken straight out of the theory of emergentism: the emerging properties of a complex system cannot be simply foretold from the available knowledge on the isolated parts of such a system, exactly as we could not say that haemoglobin has a fundamental function in carrying oxygen if we only knew the sequence of amino acids of which it is made up, but we didn't know that it flows in our blood.

Despite everything, there are excellent attempts at defining life. One of the most interesting metaphors is the one evoked by Pier Luigi Luisi in his book *The Emergence of Life* (Oxford University Press, 2006). If an alien, on a visit to Earth, were to ask a terrestrian what is living and what is not, our terrestrian would have no difficulty in separating two piles. On one side, the living: a fly, a tree, a child, a mushroom, an amoeba and a coral. On the other side, the non-living: a radio, a motor car, a robot, a crystal and a computer.

The alien, surprised by the ease and efficiency of the terrestrian, asks: but what tells the ones from the other? Movement? Certainly not: the tree and the mushroom do not move, while the robot and the motor car do. The possibility of growing? The coral does not grow much in an appreciable time, whereas the growth of a crystal is one of its most studied properties. Is it the capacity of reacting to stimuli? The mushroom and the tree do not visibly react when stimulated, whereas the computer and the radio have a number of ways of interacting with the environment around them.

In the end, the human being has an idea: they are distinguished by their ability of regenerating from inside themselves what they are made of, that is to say of forming and reproducing themselves by *autopoiesis*. Such an assumption is certainly considered interesting by the alien, although nobody on Earth has yet succeeded in creating by *autopoiesis* any system which we could define as 'living' according to our common opinion.

Other attempts at defining life would refer to such perquisites as metabolism, the ability to generate and the possibility of undergoing mutations (necessary in order for an organism to evolve). The truth of the matter is that none of the criteria evoked up to this point is sufficient, by itself, to define, in an unambiguous way, the whole of the characteristics of living beings.

Life: A Probable Phenomenon or a Coincidence?

Let us move to the second, almost impossible problem: understanding how life began. All that has been attempted up to now both in the experimental field and in the theoretical speculative one (for which I refer my readers again to the book by Ed Regis) have not arrived anywhere near solving the question. A different, more philosophical approach could be to ask ourselves: to what extent life is a probable phenomenon?

Let us try and build two scenarios: in the first one, which we shall call *deterministic*, several molecules would have assembled into more complex structures, such as nucleic acids, cells and metabolic associations pushed forward by laws of nature. Such a deterministic scenario implies that life would begin in any case, starting from different initial conditions, because the laws of nature would some-how foster life's beginning and existence.

The other scenario, which we call *contingent*, is the one in which life results not so much from chance but rather from the coincidence of several concurrent phenomena. These include the fact that homochirality is generated in life molecules, the fact that sequences of nucleic acids are created endowed with peculiar properties (such as the ability of producing self-identical copies) and the generation of proteins integrated into metabolic cycles that can support life. The contingent scenario implies that life would never have started from an even slightly different original situation.

Understanding if life started by chance or according to a preordained (or at least unavoidable) situation would allow us to better understand the emerging properties that keep a living system alive. But what is it that can tell us if life was originated according to a deterministic or a contingent scheme? Up to now, the *thermodynamic criterion of equilibrium* has been much in use. In it, the thermodynamic stability of the different molecular species which make up life is first assessed. Then, the likeliness that such species could be 'selected' by nature in a deterministic manner is established. In other words, the more a sugar, a nucleic acid or a protein is 'stable', the more it would be probable that nature may have arrived necessarily to such a molecule. In the opposite case, that is to say of non-stable molecules, the thermodynamic criterion states that only a chance coincidence could have produced them.

The thermodynamic criterion, however, can lead us into error: the homochirality of life molecules, for instance, although fundamental, would not be thermodynamics' first choice: this discipline would prefer a *racemic* state (in which the two mirror images of the same molecule are present in an equal amount). Summing up, the criterion by which to decide whether a molecular system is 'preferred' by nature's laws or whether it is the outcome of coincidence is to be selected with great care, assuming such a criterion exists.

Life: Constructive Approaches and Deductive Approaches

Another method to try and clarify how life began can be that of creating from scratch a living organism starting from inanimate matter. In several synthetic biology labs, experiments are being conducted that imply, for instance, artificially producing cellular membranes and inserting into them the ingredients of the complex molecular architecture of life. We can only hope they work. Up to this point, however, nobody has been able to put together a living organism even starting from rather complex molecules. Even if someone some day were to succeed, we could never say the way we went with this *constructive* approach is the same as the one employed at the origin of life as we know it now.

Another approach, of the opposite sign, is the *deductive* one. We start from forms of life we know and proceed as if we were to take them apart piece by piece, with the aim of cutting down their complexity. Step by step, the probability is assessed that a simpler mechanism may have evolved into a more complex one and a backward way gets traced, until the molecules from which it all began are (hopefully) found. Not even in this way, however, final answers have been obtained on how life did originate.

Life 2.0?

How many types of life are there on Earth? Up to now, only one was found, the one based on the information encoded in the genetic material made of nucleic acids. The current opinion among biologists is that it is highly improbable we should find more, but nothing forbids that, by deepening our search, we may in the end also find some other system.

Let us imagine a very young Earth, still restless but cold enough for water to exist in a liquid form. In a water solution, molecules can interact. We know that in the history of life it is exactly in water that began the transition from simple molecules to more complex ones. Here extraterrestrial amino acids could have an advantage, stemming from their experience of a difficult environment, here and there in the universe. They may have become selected for their very stability, as this fact would have allowed these molecules to move unscathed through the turbulence of the first half a billion years of the Earth's life.

Exactly at this point lies a crucial step, which is still missing in order to complete the *fil rouge* that connects Big Bang to man. For, if we roughly know how to arrive to stable amino acids, we do not know how it was possible to arrive from them to the so-called 'RNA world'. It is a kind of embryonic world, from which, later, the 'DNA world' evolved to reach the first forms of life and, in the end, also to ourselves. Given the fact that, for the time being, there is only one life, it would appear that all living organisms, ourselves included, have originated from that already living 'thing', which we fondly call LUCA (last universal common ancestor). Of course, after LUCA, during the following four billion years, life has become differentiated and complicated to a large extent, reaching today's species.

Could another LUCA exist, different from the one that originated us? In principle, yes, life might have begun more than once in different ways and maybe still exists in forms that we have not yet acknowledged.

We may not have found 'him' (or 'her') because the search for alternative forms of life on Earth has been based, up to this point, on characteristics we know, essentially because we can't think of reasonable alternatives. We have been looking for the progeny of LUCA, or for a different LUCA, based on CHNOPS, and this may be why we have found nothing different.

A different life (*Life 2.0*, in Paul Davies' words) could have different basic elements. For instance, we could imagine arsenic instead of phosphorus (i.e. not CHNOPS, but CHNOSAs). The chemical properties of the two elements are very similar to each, owing to their being *congeners,* i.e. positioned one above the other in the Mendeleev table. NASA scientists are actively looking (so far unsuccessfully) for elementary forms of life 2.0 in a Californian lake rich in arsenic.

Alternatively, we could imagine to move from carbon to silicon, again for their congener position in the table and because they both feature four potential bonds for making stable molecules (as in CH_4 or SiO_2). A first drawback of SiHNOPS (instead of CHNOPS), however, is that simple silicon compounds are in general little soluble in water—sand is not dissolved into the sea—and life without aqueous solutions appears highly improbable.

Perhaps the greatest difficulty of a possible silicon life lies in the weakness of the bonds between the silicon atoms themselves. For, if the carbon–carbon bonds are stable and strong and allow for the construction of the infinite number of the complex molecules of organic chemistry, the silicon–silicon ones are unstable and could never support such a complex sort of chemistry like the one necessary for life [palla al centro]. And, once again, thank God carbon exists and it's made like that, both as a nucleus and as an atom.

But we should not lose courage. Exactly because it is so different, life 2.0 could be very difficult to ferret out. It might nestle in places even more extreme than those beloved by extremophile bacteria, which are themselves descended from LUCA, but live happily in volcanic geysers, in gold mines, in nuclear reactors or in other generally uninhabitable places.

Nor should we be discouraged by the hypothesis that yes, maybe once upon a time there existed another form of life on Earth, but then CHNOPS, LUCA and evolution may have killed it out to take its place. On the contrary, it is by no means proven that evolution should always imply the destruction of the less successful species. Life 2.0 could be hiding in strange places and continue to exist, possibly in an elementary form, yet blooming in its isolation.

An example of shared life that has been going on at least for the past three billion years is the one between bacteria and archaea, the two kingdoms into which we now know the simplest microbes are divided. These two groups are both LUCA's children, but between them, there is an evolutionary distance not too different from the one that divides each of them from us human beings.

The only thing that remains to be said is that if 1 day we were to conclude that there is (or there has been) more than one life on Earth, this would be an important precedent for a possible generality of life, both within our planetary system and on those extrasolar planets that we are discovering at an ever-increasing pace.

Panspermia

The Greek philosopher Anaxagoras (496–428 BC) is supposed to have been the first to posit that the seeds of life had been popping out here and there all around the universe, and from there, travelling in the most fantastic ways, would then fall also on Earth. The idea (*panspermia*, all is seed) was a part of his notion of a pluralist physics made up of particles perpetually moving, somewhat similarly to the *panta rei* (everything flows) theory, posited half a century later by Democritus.

More than two millennia later, the Swedish physicist Svante Arrhenius (Nobel Prize laureate for chemistry in 1903) imagined a panspermia in which a population of ill-defined 'spores' would travel between planets, and from star to star, pushed forward by the pressure of the radiation emitted by the stars themselves. Not a bad idea, since at least it supplied a vaguely plausible mechanism for transportation. Actually Arrhenius had seen correctly also on other totally different, but very important, issues, for instance on the greenhouse effect in the Earth's atmosphere.

The interstellar panspermia idea, crazy and fascinating at the same time, was taken up in 1973 by another Nobel Prize laureate, no less a scientist than Francis Crick, together with biochemist Leslie Orgel. Crick was *the* Crick, the one who had decoded the double-helix structure of DNA with James Watson. Given Crick's (and Orgel's) fame and authority, you can imagine the raucous caused by the publication of their article entitled *Directed Panspermia*.

In it, they suggested that the passive mechanism imagined by Arrhenius for the voyage of life from star to star is something of the past. Today, it would be better to imagine that 'organisms were deliberately transmitted to the earth by intelligent beings on another planet'. The authors would give no further detail but add (thanks be to them) that the scientific proof of all this is 'for the time being inadequate' and that, in the end, still a lot of work remains to be done.

The two authors bring the idea to completion in the book entitled *Life Itself*, in which directions are given on how to enact directed panspermia. A space probe should be filled with a 'genetic starter kit', made up of different samples of resistant microorganisms that would need little nourishment; then the probe is shot into the universe to reach another planet that looks interesting, for instance, Earth.

According to Crick and Orgel, real and considerate scientists, the exogenous origin of life would finally explain some pretended chemical anomaly of life on Earth. For instance, to explain the fact that terrestrial biological systems depend on molybdenum, which is much rarer on Earth than other chemically similar metals, such as chrome and nickel. Frankly, today it seems to us to be a rather weak reason to imply that aliens (clearly rich in molybdenum) should spread their bacteria on a probe aimed at Earth. All the more unlikely because today we know the problem does not exist: in the sea (where life began), there is much more molybdenum than on dry land.

The truth is that the origin of life is a problem, as Francis Crick himself wrote in *Life Itself: Its Origin and Nature*: 'An honest man, armed with all the knowledge available to us now, could only state that in some sense, the origin of life appears at the moment to be almost a miracle, so many are the conditions which would have had to have been satisfied to get it going. But this should not be taken to imply that there are good reasons to believe that it could not have started on the earth by a perfectly reasonable sequence of fairly ordinary chemical reactions. The plain fact is that the time available was too long, the many microenvironments on the earth's surface too diverse, the various chemical possibilities too numerous and our own knowledge and imagination too feeble to allow us to be able to unravel exactly how it might or might not have happened such a long time ago, especially as we have no experimental evidence from that era to check our ideas against'.

More or less in those same years, another great English scientist wrote at length on panspermia. We are speaking of our friend Sir Fred Hoyle, the same man who supplied us with the explanation of two pieces of physics essential to our life: stellar nucleosynthesis and the 'privileged' structure of the carbon nucleus. Fred Hoyle did not win a Nobel Prize purely by reason of a spectacular piece of injustice.

In his book *The Lifecloud*, Sir Fred affirms that the process of formation of big organic molecules within the interstellar medium, the so-called bricks of life, appears to have gone well beyond what we can imagine. On planetesimals and asteroids, but mainly on comets, life would have formed a bit everywhere (probably because of panspermia), although in an elementary form. Life would then have been brought on Earth mainly by comets. In another slightly later book, *Diseases from Space*, Fred Hoyle goes so far as to attribute to comets not only the origins of medieval plague outbursts but also the continuous diffusion on Earth of germs only slightly more benign, such as the viruses of influenza. Statistics in hand, Hoyle associates the contemporary explosion of flu epidemics in English schools, set hundreds of miles apart, to the unavoidable—for him—fall out of space of viruses from a comet. Poor English schoolchildren!

But there's more to come. The presumed fall from the sky of forms of life, dangerous ones included, moved Hoyle to develop an interesting theory on the anatomy and physiology of the nose development in human's existence. At the dawn of human existence, the great scientist tells us literally, their noses appeared to be turned more upwards than today and thus all forms of life could fall into them, including those sprinkled by passing comets. For this reason, little by little, our noses would have evolved into their present shape, turned downwards and thus much more protected. Hoyle was really a very great scientist, but with the story of our noses, he may have gambled away his Nobel Prize.

Today, nobody credits panspermia as a viable theory any longer. And we, bereft of ideas on the origin of life, cannot but wonder once again: if we are the Martians, who are the (real) Martians and where do they come from? From another nameless planet and so on to infinity? The answer is not bound to arrive very soon. However, a hand will be lent to us by mathematics and computer science that today are making biology bloom as the case was with physics in the first half of the twentieth century. Perhaps, thanks to the new research field of system biology, we will no longer be obliged to call upon an extraterrestrial *deus ex machina*.

It's a pity, in a way, not only because panspermia is an amusing, albeit fantastic theory, but also because—in one of those paradoxes that make science fascinating—it has recently been demonstrated that interplanetary (or maybe interstellar) voyages of living microorganisms are possible: we have thus put an end to one of the traditional issues used *against* panspermia.

In a recent experiment conducted on the International Space Station, with Italian industries and scientists also taking part, colonies of living microorganisms have been exposed on trays outside the Station for many months. Surprisingly, the protracted exposure to an extreme thermal environment, to void, to powerful radiation and to absence of gravity did not kill all the microorganisms. Back on Earth, after being administered a few drops of water, many of them returned as lively as ever.

This experiment tells that interplanetary panspermia could be not impossible; this could be all the more true if we think that meteorites from Mars (as well as from other planets) are able to offer in their nooks and crannies a better shelter than the tray set outside the Space Station. Summing up, life as we know it could survive in space even for a long time, contrary to what was current opinion till a few years ago.

What that scatterbrained genius of Sir Fred Hoyle would say, had he known, as we know today, that many comets of the solar system were generated around other stars. And maybe Francis Crick and Leslie Orgel would immediately have refreshed their model of guided panspermia, being able to use a fantastic interstellar vector such as comets, rich in water and organic material, more welcoming than any probe for the voyage of microorganisms sent by intelligent aliens.

Chapter 8
Is There Anybody Out There?

Until very recently, it was taken for granted that there would be somebody else out there, not just elementary forms of life, such as the bacteria that had lorded over Earth for billions of years (and which today still make up the bulk of our biomass). No, when we said 'somebody', we meant developed and intelligent life.

Doubts began in recent decades, when we have been able to really start looking for that 'somebody'. For up to now, our searches of the sky have revealed nobody at all. We found only life's basic building bricks. That's a lot—a major discovery—but it's not life.

But let's try looking at it this way. We have seen these building blocks, i.e., big organic molecules such as amino acids, out there in celestial objects, and we have found them in meteorites a few decades after 'discovering' such molecules here on Earth, that is to say, shortly after science had understood what they were and what their significance was. Similarly, we acquired an idea about how the 'bricks of matter' (the nuclei of heavy atoms, produced by stellar nucleosynthesis) were formed and distributed in the universe of only a few decades after understanding their physics on Earth. Both time intervals were mighty short, even on the scale of human evolution, to say nothing of the cosmic one.

There's one obvious conclusion: let's give ourselves time, let's not get discouraged, let's keep studying on Earth and researching in the sky.

But let's proceed in an orderly way. It would be difficult—and fruitless—to recount here the full history of human thought and emotions concerning the notion of extraterrestrial life. Such a history

G.F. Bignami, *We are the Martians*,
DOI 10.1007/978-88-470-2466-3_8, © Springer-Verlag Italia 2012

is a mix of sublime poetry and silly fancies, of yearnings and deep terrors, of pseudoscience and ignorance, in short, of projections of ourselves into the outer world.

We shall only refer again to Giordano Bruno, the sixteenth-century thinker who believed in the existence of infinite, inhabited worlds. He was not burnt at the stake just for this but for other 'heresies' that the Church (of Rome) in those times deemed to be far more serious. However, the idea itself of inhabited worlds was heresy to a clergy still unable to face the idea that there might be a possibility of life outside Earth.

Less than three centuries later, however, in France—itself a Catholic country—Camille Flammarion could publish without fear his book *Les Terres du Ciel*, a beautiful volume of popular astronomy, complete with a fanciful census of the inhabitants of the solar system. Today, the Catholic Church has opened up even further. In 2009, the Vatican Observatory itself gave a greeting to 'our extraterrestrial brothers', if they exist.

To illustrate a more recent past, we shall move beyond the last tumultuous century of observations (and figments of imagination) to a very special case: the planet Mars. To speak of the future, we shall instead investigate the means by which we can seek intelligent life outside the solar system, since by now it appears as a fact that we shall not find it in our immediate neighbourhood.

The Saga of the Martians

The legend of aliens from Mars began a little more than a century ago, in Italy. It then spread to the United States and from there to the whole world.

It all started with Giovanni Virginio Schiaparelli (1835–1910), a great astronomer who was the head of the Brera Observatory in Milan from 1862 to 1900. In 1877, while observing with his new German telescope (the 218 mm Merz), Schiaparelli fell in love with Mars and began drawing its surface. In those days, you had to position your eye directly onto the eyepiece of the telescope—without the help of the photographic or television cameras we have today. In the rare moments in which the image was clear, you had to memorize what

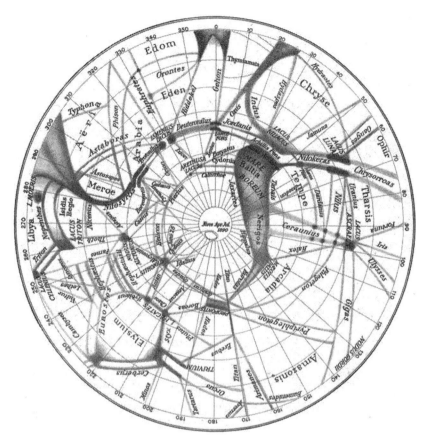

Fig. 8.1 Mars's boreal hemisphere drawn by Schiaparelli, based on observations made in 1890. (Atti della Reale Accademia dei Lincei—Memorie Cl. Sc. Fis., mat. E nat. , Serie V, vol. VIII)

you saw—or imagined you saw—and then draw the image by hand, maybe in the cold and by uncertain candlelight.

Taking advantage of what Flammarion would at the time call 'le ciel limpide et calme de Milan', Schiaparelli discovered and drew the most important elements of Martian topography, explaining the chromatic diversity on the planet's surface with the presence of seas and continents (Fig. 8.1). In his eyes, some 'seas' appeared to be interconnected by means of what he called by the Italian word 'canali', a fateful choice. At first, he had not meant to say they had an artificial origin: the same word is used for narrow sea branches or straits, such as the English Channel or the Sicily Channel.

Schiaparelli's results appeared in the *Rendiconti* (Proceedings) of the Accademia dei Lincei in Italian, a language still widely read around the science world at the end of the nineteenth century. The international success of his work was immense; so much so that, despite publishing not a single word in English, he was awarded in 1872 the golden medal of the Royal Astronomical Society, as well as, in 1902, the Bruce Medal, the highest recognition of the American Astronomical Society. No other Italian astronomer has received either of the two medals since.

Schiaparelli's images and prose charmed a rich American diplomat, Percival Lowell, born in Boston in 1855. Lowell was so attracted by astronomy that he gave up his diplomatic career and invested all his (remarkable) fortune in the building of a new, beautiful observatory in the mountains of Arizona. The Lowell Observatory is still operational today.

Before he began observing Mars, Lowell had read about Schiaparelli's 'canali', a word which he did not translate into English as 'channels' (as in the natural sea way) but as 'canals', phonetically very similar to the Italian one, but which in English implies an artificial origin (such as the Suez or the Panama Canal). The mistake is not surprising at a time when human progress was marked by the very digging of great canals, and in a general context, that tended to yearn towards believing in the existence of intelligent life on Mars.

Lowell 'saw' on the planet a real cobweb of canals, very long and very narrow, which he described as 'practically monodimensional'. He made a Martian globe on which the canals were arcs of circles and would intersect by two or three—or even by seven—points. It all looked very much like cities served by a complex network of waterways.

At this point, Schiaparelli also seemed to give way to imagination, albeit with a certain restraint. In 1895, he published an amusing article entitled 'Life on Mars' (La vita su Marte). On his own copy, which is held in the Historical Archive of the Brera Observatory, Schiaparelli hand-wrote an enlightening epigraph, *Semel in anno licet insanire* ('Once a year one can go foolish').

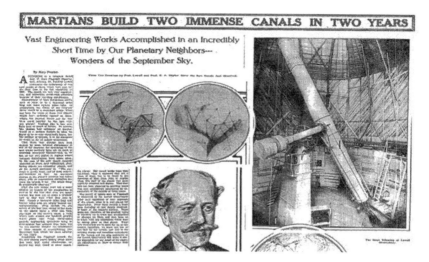

Fig. 8.2 Percival Lowell and his famous 'discoveries' as reported by the *New York Times* of 27 August 1911

The text said that on Mars, there exists a central organization for the transfer of waters, under the orders of the Great Prefect of Agriculture. At the time of the melting of polar snows, the Prefect orders the opening of the locks in order to make water flow into the complex systems of canals for watering the countryside. So, wrote Schiaparelli, Mars 'is for sure the paradise of hydraulic engineers!' Indeed, the obvious common interests of its inhabitants turn Mars also into 'a paradise for socialists, aiming at fighting together against the common enemy, that is to say difficulties raised by the meanness of Nature'. Talk about *insanire*.

After Schiaparelli's death in 1910, Lowell was left undisputed master of the field and, with no one to put a damper on his enthusiasm, he went so far as to 'demonstrate' that Martians, by now generally accepted to exist, were in actual fact remarkable engineers. Thanks also to his social standing, in 1911, Lowell managed to have the *New York Times* announce on a full-page title (Fig. 8.2) that 'Martians build two immense canals in 2 years'.

Thus, a century ago, everybody believed not only that there was life on Mars but also that Martians were intelligent and more capable than ourselves in engineering works. And all this because of a linguistic blunder that—between Milan and Arizona—had first waylaid two great astronomers, then journalists and then a vast public the world over.

Lowell died in 1916, when he was only 61, and is buried on a hill— 'Mars Hill'—near his telescope. His premature death spared him the pain of seeing his Martian visions dissolve. Astronomic photography was about to arrive on the scene, demonstrating clearly that the surface of Mars was not at all the way Lowell had imagined it.

Yet the myth of the Martians was hard to kill, especially in the public mind. Edgar Rice Burroughs (1875–1950), the creator of Tarzan, wrote 11 books within the space of three decades whose protagonists were Martian princesses and Martian warriors, and they were immensely successful up to the 1940s.

But the most spectacular example of mass hysteria on a Martian theme was Orson Wells's celebrated radio broadcast 'The War of the Worlds', loosely drawn from the book by H.G. Wells by the same title. It went on air on the evening of October 30, 1938. It was a spoof radio report of the bloody invasion by Martian spaceships of New Jersey. It had an unanticipated impact, though, as listeners were completely taken in. Panic was unleashed. The roads and the telephone systems jammed, and the police and the army had to step in. Judging from the success of today's television broadcasts for gullible viewers on 'mysteries of the universe', maybe a similar panic might spread today too. A spoof special edition of TV news would perhaps suffice where it showed tin spaceships with a licence plate reading MARS and little green men on board.

It actually took decades to prove that artificial canals on Mars did not exist. In 1965, the great planetologist and science fiction writer, Carl Sagan, was finally able to write in *Nature* that Mars has neither canals nor oceans and certainly doesn't have princesses or launch ramps for spaceships. The nature of the 'canals', he said, is simply geological or volcanic. Sagan could affirm this because, a few months earlier, the flights of fancy had clashed against the first photographs from the red planet, taken from the probe Mariner 4, that showed a desert, almost moon-like landscape.

No, there are no developed and intelligent beings on Mars, much less so on other planets of the solar system. The possibility may remain of finding forms of elementary or fossil Martian life, such as the famous 'wormlet' on meteorite ALH84001 (see Chap. 4).

Let's accept it: it's us who are the Martians, even though, in our collective imagination, Martians—the 'real' ones—are, and will always be, something alien, something else, even if, or just because, they do not exist.

Ozma and Drake's Equation

It is not realistic to hope for a physical contact with life outside the solar system, not even in our galactic neighbourhood. The few light years that separate us even from the closest stars are unremarkable distances on a cosmic scale but well out of the reach of any voyage of exploration, much less one with a crew on board. The impossibility of visiting aliens on site, however, should not discourage us—because 50 years ago two visionary physicists set the way for searching a contact with extraterrestrial life by means of electromagnetic waves.

Giuseppe Cocconi (1914–2008) and Philip Morrison (1915–2005) were already well-known physicists when, in September 1959, they published an article in *Nature* entitled 'Searching for Interstellar Communications', pointing to a simple way: all you have to do is to tune in on the most appropriate radio frequency and patiently wait.

Cocconi, who had started experimenting with Enrico Fermi, went on to a brilliant career at CERN in Geneva. Morrison, professor at MIT in Boston, had been the head of a group in the Manhattan Project. Their letter to *Nature* began by setting out the obvious. At the time, there was no evidence of the existence of planets around other stars, there was no idea of how life could emerge on them nor of how technological societies could develop on them. In any case, they said, if intelligent beings were really to be found some place up there, they might have created a system of communication addressed to the rest of the universe. If so, we might make use of that system ourselves.

Radio waves are the most efficient way of transmitting a signal over distance. Cocconi and Morrison suggested that the new antennas for radio astronomy—which was being developed in those days—should be employed for listening in on 'universal' frequencies, such as the one characteristic of the physical structure of the hydrogen atom. Of course they had no idea of what signal to expect. Sequences of prime numbers? Digits of pi? No use trying to guess, they said, let's trust *them*.

The suggestion of the two physicists produced a lot of enthusiasm. In April 1960, Frank Drake, at the US National Radio Astronomy Observatory (NRAO), which had just been created in Virginia, started the Ozma Project, the first research for an intelligent cosmic signal in radio waves. Ozma is the queen of the country of Oz, born of L. Frank Baum's imagination and made famous by the film *The Wizard of Oz*. But Drake did not intend to step into a magic world. Quite the contrary, he had very clear ideas on the enterprise he was entering into. In order to assess his probability of success, he wrote a formula which later became famous.

The aim of *Drake's equation* was to give an estimate of the number N of civilizations in our galaxy capable of sending radio signals that we might receive. Here it is:

$N = R$

But what do the second member factors mean?

R is the rate at which those stars are formed that can allow, in principle, the appearance of life ('habstars').

f_p is the fraction of those stars that have planetary systems.

n_e is the fraction of those planets which have favourable conditions for life (in the 'habzone').

f_l is the fraction of such planets where life in fact develops.

f_i is the fraction of the planets in which, once life has developed, intelligent civilizations arise.

f_c is the fraction of civilizations that develops a technology capable of sending radio signals into space.

L is the amount of time covered by the sending of the signals.

In order to calculate the probability of a contact, that is, the number of civilizations sufficiently advanced that are present in the galaxy today, it is necessary to multiply the rate of habstar formation times the fraction of these with planetary systems, times the fraction of planets positioned in a habzone, times the fraction of these where life has in fact developed, times the fraction where life has generated intelligent civilizations, times the further fraction in which civilizations acquire a technology capable of sending radio signals into space and, finally, times the amount of time (expressed as a fraction of the life of the universe) during which such signals have been emitted.

Some factors in the equation we know only in a very partial way but most of them we do not know at all. Drake's equation serves then to summarize the problem, but it is not of much help in giving a true estimate of the probability of establishing a contact with an extraterrestrial civilization.

But 50 years did not go by in vain. Today, we have a much more exact idea of the rate of formation of habstars and of the probability that a planet could be in the habzone, i.e., be (remotely) compatible with life as we mean it. It's a step forward, but we're still far from being able to calculate the number N of civilizations in the Milky Way which are active today. Indeed, we are still far from understanding if a single one exists.

SETI: Galactic Eavesdropping @ Home

Let's go back to 1960: young Drake, just arrived to work with the Green Bank radio telescope of NRAO, reckoned that the instruments at his disposal could reveal the presence of radio signals coming from an intelligent civilization at the maximum distance of 10–15 light years from us. Thus, he decided to concentrate his efforts on listening in to two stars very similar to the sun—Tau Ceti and Epsilon Eridani which are, respectively, 12 and 10.5 light years away. Note that in 2006, it was also discovered that Epsilon Eridani has one planet (and possibly two). Thus, at least one of the targets of the very first research into extraterrestrial intelligence had been well chosen.

Drake kept this activity of his a secret, to avoid the risk of being considered a visionary lunatic. Yet he had no reason to worry: for, precisely in 1960, NASA would start a line of research in *exobiology*, aimed at the study of possible forms of life in space. Since then, more than 100 programmes of search and listening have been carried out; among these, the most famous one is no doubt SETI (*Search for Extraterrestrial Intelligence*), which NASA started to finance in 1971.

Although SETI is a programme dedicated mainly to listening for possible celestial signals not attributable to natural causes, SETI has also *sent* messages towards the universe. In 1974, for example, the radio telescope in Arecibo (Puerto Rico) was used to send a coded signal towards the globular cluster M13. The message was a 23×73 matrix (two prime numbers) in a binary format; the sequences of 0 and 1 representing numbers from 1 to 10, the atomic numbers of the most important chemical elements for life on Earth, the chemical formula for nucleotides, the double-helix structure of DNA, a few data on the solar system, a drawing of a man and of a radio telescope, complete with their respective measures. It will in any case take 25,000 years for the signal to reach M13. Let's not expect an immediate response.

SETI was financed by NASA until 1993, when the lack of results caused funds to dry up. Left without federal support, the authors of the project have been appealing for financing by the general public. SETI has since lived on donations by private persons willing to invest in a dream.

In order to ensure the survival of the project, SETI researchers have developed an extraordinary ability to adapt. First, they have learned to conduct radio observations at no cost, by 'parasiting' other people's observation time: while the radio telescope is collecting data on any possible celestial source, the SETI receiver works its own independent way.

Secondly, a zero-cost system was invented for data analysis. The huge computing power necessary to analyse all the radio signals collected from the sky is supplied for free by a network of millions of personal computers thanks to the system of *distributed computing* called SETI@home. By connecting with the site setiathome.berkeley. edu, anybody can download the project software which, once installed, starts operating when the computer is on but not working: thus our own PC, in its spare time, analyses data that no one has yet seen.

This means that any one of us, some day, could be the first to discover an extraterrestrial signal. For many, that's an irresistible prospect, so much so that, since 1995, above five million people have already downloaded the programme. In this way, the hope of intercepting a galactic signal has also generated an extraordinary experiment of 'collective computing', the first in the history of computer sociology.

Still, nothing has been found. Does that mean we are alone in the universe? Not at all. As Francis Bacon wrote: *'they are ill discoverers that think there is no land when they can see nothing but the sea'*. And Jill Tarter of the SETI Institute makes use of another effective metaphor: deducing the absence of intelligent life in the universe from the null (up to now) result of SETI would be like denying the existence of fish in the oceans after collecting a glassful of sea water and not finding fish in it. Yes, we could have been very lucky and caught a tiny fish, but it is infinitely more probable that the first glassful should contain no fish. If we are to nourish a hope of success, the glassful must become a bucketful, then a tubful and so on.

In this sense, SETI has been a very useful experience. Also thanks to this project, in the past 50 years, our ability to look for radio signals has improved *10,000 times more* than the sensitivity of optical astronomy has improved in the four centuries between Galileo and ourselves.

The 'Berlusconi Bubble'

We have no idea of how forms of extraterrestrial life might manage to send radio signals in space. To analyse the issue, we can avail ourselves of a sample made up by only one planet, our own. We know that, seen from the outside, our Earth is surrounded by a sphere of electromagnetic waves that expands at the speed of light in all directions. In the century since Marconi began to broadcast by means of radio waves, this sphere has engulfed all the numerous stars that are within 100 light years from us. The signal has very much intensified in the past few decades: in Italy, we can fondly call it the 'Berlusconi bubble'. For sure, one could argue whether television advertisements, talk shows and such like make the ideal message to be broadcasted for presenting our civilization to intelligent aliens.

Our ability to send signals in the frequency band of visible light has also improved remarkably. Commenting on the 50th anniversary of our so-far unsuccessful attempts to exchange messages with our galactic neighbours, in 2010, physicist Paul Horowitz remarked that it might be better to abandon the radio channel in favour of emission from pulsating laser beams.

With the most powerful lasers available today, our Earth could be made to shine like 10,000 suns, although in tiny flashes of a few billionths of a second—a 'dazzling' strategy that could render us visible to possible inhabitants of planets belonging to other stars, assuming they look at their night sky with due attention.

Frank Drake is convinced that SETI represents a search for ourselves and of our place in the universe. There's no doubt that the discovery of some form of life outside Earth, intelligent or otherwise, would represent a philosophical revolution of extraordinary significance. We would perceive we are only one of many possible life forms. It would also be the last and final blow to anthropocentrism.

Cocconi and Morrison concluded their letter to Nature by writing (in these very words):

'The probability of success is difficult to estimate; but if we never search, the chance of success is zero'.

A Message in a Bottle

Meanwhile, we terrestrials didn't only send photons around. We have already dispatched solid, cut-metal messengers into the depth of space. They are four probes, Pioneer 10 and 11, plus Voyager 1 and 2, sent off more than 30 years ago. All four objects, built by human beings, are today leaving our planetary system, as well as the zone influenced by the sun, and are already entering the interstellar medium. We have entrusted important messages to them.

The two Pioneers carry the gold plate shown in Fig. 8.3, designed by the creative Carl Sagan, together with Frank Drake. They tried to concentrate within it most pieces of information potentially understandable by an extraterrestrial civilization. Engraved on it, there is the scheme of a fundamental property of a hydrogen atom (the so-called hyperfine structure), a scheme of the solar system complete with the probe trajectory, the position of the sun relative to 14 pulsars (a very precise way of pointing out our astronomical position), a drawing of a man and a woman standing next to the spacecraft (to give an idea of the size of our species), and a binary code.

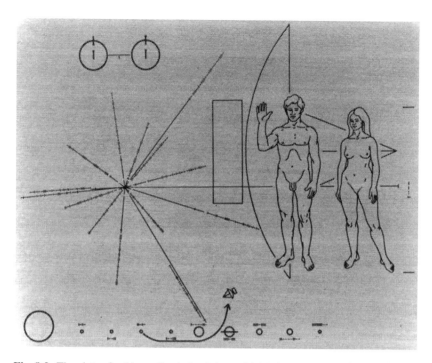

Fig. 8.3 The plate of gold-anodized aluminium which left the solar system on board the Pioneer probes 10 and 11. Photo. NASA

There is something similar, but even more elaborate on the two Voyager probes. Besides the schemes of the position of the sun and of the hydrogen atom, each of the probes carries a gold 'sound' record (in 1977, at the time of the launch, the audio CD had not been invented). Complete with directions for use, it contains especially significant recordings of terrestrial and human sounds, from the whispering of the wind to a baby crying to the masterful ta-ta-ta-taa of Beethoven fifth symphony. It is hoped that someday, someone may find the record, may listen to it, may get an idea of what life on Earth is like and may then choose to go.

Chapter 9
CODA: What Remains to Be Discovered?

Almost everything, we might say: we can certainly not rest on our laurels. True, we believe we understand something about fossil radiation, nucleosynthesis, the birth of the Earth and of the extrasolar planets, interstellar molecules and DNA, about evolution and lots more. We have discovered that we are directly linked to the Big Bang, whose traces are still within us, that we are made of stardust and that in outer space, there is an abundance of the 'bricks of life', though we have not yet found anything alive out there.

But the thread that we have tried to follow, far from being continuous, shows tears and gaps. In some spots, it's missing altogether, leaving us to deal with unsettling 'black holes' in our current knowledge.

There is a black hole right at the start: Plank's time at the beginning of the universe. It is a very short but significant interval (10–43 s), of which we know nothing. We do not even know where to start filling in this gap. We could imagine two possibilities: a 'theory of everything', which expands today's physics putting together all the properties needed for the birth of the universe or a 'new physics', which would lead to the same result, starting from assumptions yet to be invented. But these are only words; it is like whistling in the dark.

The other abrupt interruption in the Big Bang-to-man thread has to do with life itself. Here we have not one, but two black holes of knowledge. They are tightly bound together (like in a binary system in the sky), so that if we were to solve one, we might find the way to get to the other one.

G.F. Bignami, *We are the Martians*,
DOI 10.1007/978-88-470-2466-3_9, © Springer-Verlag Italia 2012

One knowledge hole refers to the origin of life on Earth. As we have seen, we're talking of the origin of something we do not even know how to define, if not in a partial or insufficient manner. We have identified and can handle the bricks, but we do not yet know how to build the house.

The second black hole refers to the presence of life elsewhere in the universe. At first sight, it appears to be a simpler problem or at least a better defined one: if we were find 'life' (with astronomical instruments yet to be invented), we should be able to recognize it. But is that really so?

Maybe it is only our limited imagination that makes us believe that one day, if we won't see a three-legged chicken scratching about on Mars, at least we will find traces of chlorophyll on a planet's spectrum or a string of digits of Pi in a radio transmission from a planet around Alpha Centauri. Too simplistic? Maybe. The fact remains that we have found nothing of the kind yet; it could be that we are still just whistling in the dark, without knowing what to look for.

In any case, 'we are the Martians'. However, we still do not know if it is so because we truly come from far away or because we are really so unique that, for lack of better actors, we must to play the part of the aliens as well. . ...

Those three knowledge black holes exist because, to be filled, they would require stuff that we do not have today and cannot even foresee. Luckily, however, amongst the infinite things still to be discovered, there are many for which we have at least some vague idea on how to proceed. Let's look at these knowledge gaps as if arranged on a series of shades of grey.

Sir John Maddox, director of Nature magazine for 23 years, rightly said that the future does not usually bring discoveries that we expect but those totally unforeseen. A scientist in the early 1900—at the dawn of modern physics, astronomy and biology—would certainly not have expected general relativity, quantum mechanics, the Big Bang cosmology, nucleosynthesis, DNA, the mapping of the human genome and all the rest.

Today, however, we are still faced with a problem that has been looming unresolved for a century: that of the physical structure of space-time and of the matter within it. In the mid-1800, electromagnetism seemed solved by

Maxwell's equations and by the hypothesis of the existence of an 'aether' which allowed electromagnetic waves to propagate. Later, the results of Michelson's and Morley's experiment showed that the cosmic aether does not exist, throwing many eminent physicists into a panic. Something similar is happening today to cosmologists, after their discovery that dark matter prevails over ordinary one.

The problem of dark matter—does it really exist? How much is there? And, above all, what is it made of?—is certainly today another grey area, actually dark grey along our thread. To tackle this problem, we know we need space missions to give us a more complete picture of the distribution of matter in the universe. And we know we'll move forward when we'll learn how to put together results from astronomy and from fundamental physics like those out of CERN in Geneva. Maybe we will end up finding something that will make things even more complex, but in the next 20 or 30 years, there will be discoveries. That's just the time it will take for today's children to grow up, to study, to imagine the science and the right tools for it and to build and deploy them. At the end of the road, there could be a Nobel Prize: good luck!

But present and future scientists do not just live and work—luckily—for Nobel prizes. Astronomers too have their share of grey areas which await new discoveries. An example of this, much closer to us, is the understanding of the origin of the solar system.

Here we know that key elements may be hiding inside comets. The famous Halley comet, which comes close to Earth every 76 years, has been with us for a long time and will stay around much longer, even if sooner or later it will disappear, due to the numerous rotations around the Sun. Besides acting as cosmic calendar, this comet has served in the past as a gauge for mankind's scientific and technological development.

For centuries, Halley's best image has been the one painted by Giotto, after the passage of 1301; all its other renderings were mediocre or bizarre. Then, in 1910, astronomers were able to photograph the comet, to record its real appearance. With due respect for Giotto's talent, this was a great scientific advance. During the following Earth fly-by, in 1986, we were even able to examine it with a space probe that went in as close as a few 100 km from the nucleus.

Halley will faithfully return in 2062. What will we have invented in the meantime? What will future astronomers do to Halley then? They are today's new-born kids, and have little brothers and sisters who maybe future astronauts bound for Mars?

Maybe before that time, we will have found out that it may be worthwhile capturing the comet, because it contains something very interesting or useful, and maybe we will even be capable of gently guide it towards the Earth. Or, better still, maybe we will have learned that it is better to leave comets alone, shining in the sky in all their beauty. But in the meantime, we might have learnt from them the secret of the birth of planetary systems.

Another grey problem par excellence is for the coming generation of astronomers to disprove the second provocatory law of Tremaine, which states that 'planets formation is impossible'. We might even make it before the return of Halley, because in the meantime, we will have discovered and studied millions of other planets, including who knows how many Earths (the first of which, I bet, will be identified very soon, for sure within the next decade).

Finding new clues by studying extrasolar planets, particularly the 'right' ones, would be of great help in our understanding of the origin of life. Waiting for that to happen, it is necessary to continue here on Earth, starting with prebiotic molecules. For example, which role might have been played by volcanic sources at the bottom of our oceans? Reactions from iron sulphates and hydrogen from the ocean's crust create molecular hydrogen, which has the power to reduce carbon monoxide. This could be a starting point for the creation of ever more complex organic molecules within Earth's oceans (and maybe on other planets as well). As we mentioned, the 'blue blood' of some terrestrial arachnids and the excess molybdenum in our biology are amongst the numerous evidences that life on Earth started indeed in the water.

Then, there is the problem of amino acids. We have observed them in space, and we know that they are constantly bombarding us, carried by meteorites. Why and how did they come to exist? Why just those? Is it correct to start from glycine, the simplest of them all, to try to understand more complex ones? And why are we made of those specific 20 amino acids, while there are hundreds of others spread all over the universe and the solar system? The answers to these questions would also be worthy of a Nobel prize.

Another bottleneck on the path to the origin of life is the natural polymers that store and transmit information necessary to all life forms on Earth: when and how have those long and delicate molecular chains come into existence? And when and how did they start to function independently?

The discussion on large organic molecules and amino acids, a crucial point for our understanding of the BB-t-M thread, leads us to the last and most spectacular example of grey area, in this case as dark as London's smog. We're talking of biological homochirality, a big word that means that in nature, there are inexplicable preferences for certain types of biomolecules instead of their mirror images. Our amino acids are preferably of type L (levorotatory), while biological sugars are of type R (right rotatory), in a kind of biochemical equal opportunity policy. A priori, each configuration would be equally probable, and we would thus expect to find them in equal proportions in living organisms. But this is not so: owing to a strange kind of magic, our mirror image might not be good for life.

The source of this magic is yet to be explained. Maybe we will be able to explain before the return of Halley's comet, especially if biology and chemistry will join forces with physics. Maybe, for example, the asymmetry of the blocks of life will be shown to be related to the so-called 'parity violation' of particle physics and maybe even with the equally mysterious asymmetry of matter and antimatter. In that case, life would be the natural product of an asymmetric universe: not bad for a discovery.

In the present short digest, we have not touched upon biological evolution, that is, on how life on Earth has changed in the span of billions of years, in passing generating man, a special primate distinguishable from apes for two missing chromosomes. Oranguntangs, gorillas and chimpanzees have 24 chromosome pairs; in our case, one pair, who knows why, has become joined to the existing number two pair. It is a mutation that we somehow feel when we stare deep in the eyes of one of those apes, our close relatives. A simple genetic mutation, like innumerable others, but one with far reaching consequences for us. Lucy was born this way, the small African lady hominid from whom we all come from, much in the same way as all forms of life on Earth come from LUCA.

We have not taken up this subject in part because we already know we won't find on Mars animals to glare at. Also, we have not discussed that world of ours, invisible yet so important: the world within our heads. But we cannot ask too much of a reader's digest of the universe. Hence, we have set our boundaries to the world outside.

We will leave the last word to Immanuel Kant, who intuited that to make a world like our own, you just need matter from the stars. He concluded his Critique of Practical Reason with the following:

> 'Two things fill the mind with ever new and increasing admiration andawe, the oftener and the more steadily we reflect on them: thestarry heavens above and the moral law within'.

Chapter 10
15 Stories To Get Rid Of

1. The Universe has always been in existence.
The universe as we know it had its origin from Big Bang, in which the matter also was formed of which we are made. Basing themselves on different astronomic measurements and on Einstein's theory of general relativity, cosmologists nowadays believe that Big Bang took place some 13.7 billion years ago.

2. The Universe is infinitely big.
The Universe is very big and astronomical observations point towards its continuing expansion. Its size however is finite and it is determined by the amount of matter and energy that it contains inside itself.

3. The atoms we are made of appeared at the time of Big Bang.
In the first 3 min after Big Bang only the nuclei of the lightest atoms were formed, hydrogen and hellion: this was the phase of the so-called *cosmological nucleosynthesis*. The nuclei of the other heavier atoms, which make up the matter of our daily life, were generated, on the contrary, by successive generations of stars, by means of *stellar nucleosynthesis* (the result of nuclear fusion) and with supernova explosions.

G.F. Bignami, *We are the Martians*,
DOI 10.1007/978-88-470-2466-3_10, © Springer-Verlag Italia 2012

4. **All matter in the universe is similar to the one we are made of**.

In Big Bang "antimatter" was also formed with special properties which are the mirror image of those of ordinary "matter". In the two cases the particles are identical, but have an opposed electric charge, and if they meet they annihilate each other by turning mass into energy. Somewhere in the universe there might exist stars or galaxies made of antimatter that have never come in contact with matter of our type.

5. **Stars can have any dimension**.

There is a lower limit to the size of stars: an interstellar cloud principally made of hydrogen in order to be able to "collapse" and start the nuclear fusion reactions must have a mass of at least 10^{29} k. Probably there exists also an upper limit: the biggest stars known have a mass a few hundred times bigger than our own Sun.

6. **Stars are eternal**.

Every star is born, shines for a certain amount of time and then dies, either gradually going out or exploding as a supernova. Every star's life depends on its mass when it is formed: more massive stars emit more energy, quickly using up their "nuclear fuel," so have a shorter life.

7. **Our Solar System, with many planets revolving around one central star, is a unique case in the universe**.

Thanks to recent astronomical observations we know hundreds of other planetary systems set around stars similar to the Sun and we are discovering more and more new ones. So it is by now clear that for a star like the Sun having planets is the rule, not an exception.

8. **Thanks to science we know today how the Earth was formed**.

The study of terrestrial rocks and of meteorites has allowed us to establish precisely *when* our planet was formed: some 4.67 billion years ago.

On the contrary, we not yet have a detailed understanding of *how* planets are formed starting from the disc of dust that surrounds a star just born. Uncertainties on this point have increased with the discovery of extra-solar planetary systems: their characteristics are often very different from those of the Solar System, something which is compelling scientists to go back on their theoretical models of planetary formation.

9. **Surely there exist extra-solar planets that harbour life**.
It is possible and many think it probable, but it is not by all means certain. For, around a star life as we know it can only develop in a limited "habitable zone" and only a small portion of extra-solar planets will be found by chance in that zone. Moreover, to be able to harbour life, a planet must have further special characteristics, which are not necessarily frequent: a mass sufficient to hold an atmosphere, a magnetic field that can screen cosmic ionizing rays, and a tectonic structure to recycle the elements of the superficial crust.

10. **There is no life outside Earth.**
We have no evidence as yet that life may have originated also outside our planet. Yet some recent discoveries—the presence of amino acids in meteorites and in comets, the existence of a very large number of extra-solar planets—make more probable the fact that more biological matter may exist in places different from the ones we know.

11. **No living being can survive if exposed to the vacuum of interplanetary space.**
The environment of space is certainly very hostile: it lacks air, water, nourishment, and one is exposed to extreme thermic excursions and to radiations of all sorts. Yet among the microorganisms there exist very tough bacteria: a recent experiment conducted on the International Space Station has shown that they can survive after an exposure for months in outer space.
Their spores, hidden for instance in the inside cracks of a meteorite, could in theory survive long voyages in space and thus transmit life from one planet to the other.

12. **In the future humankind will certainly colonize planets belonging to other stars**.
There is a small problem: cosmic distances are really gigantic. The longest voyages done up to now by humankind in

interplanetary space (to the Moon and back with the Apollo missions) lasted less than a fortnight. With the propulsion systems now either in existence or imaginable, months would be needed merely to reach the planet Mars and to reach the closest star to the Sun one should travel for tens of thousands of years. Humankind may never be able to develop such technology as will allow them to perform interstellar voyages.

13. **There have already been contacts with alien civilizations**.

It is by now sure that in the Solar System there are no forms of intelligent extraterrestrial life, nor does any evidence exist of contacts between us and alien civilizations from extra-solar planets. If those civilizations exist in other parts of our Galaxy, it would not be impossible, in principle, to communicate with them via either radio or light signals. The huge distances and the necessity of finding a common language, however, would make this feat very difficult.

14. **Aliens have left traces of their descent on Earth**.

The arrival on Earth of aliens coming from planets that revolve around another star are very improbable, because, as we have seen, the voyage—at least according to our present knowledge— would take a very long time.

There are some who spoke of the presence of aliens to explain the building of impressive monuments of some human civilization belonging to the past; but these people show they ignore the ability, the determination and strong-mindedness of our ancestors.

15. **Science has nowadays solved the mystery of life**.

Even the most recent achievements in biotechnologies are based either on the connection of pre-existing biological elements or on the synthetic imitation of structures of human beings to be observed in nature.

Up to now nobody has been able to create a living organism starting from inorganic substances and nobody can explain in a convincing way how life originated on Earth.

In fact there is more: scientists have not even yet found a clear and univocal definition of what is life.

Chapter 11
You Might Not Know That...

We have a 'photograph' of the universe when it was very young.
The discovery of 'fossil radiation', a residual of Big Bang, which can still be measured today, has allowed us to visualize how matter from primordial cosmos was slightly wrinkled by tiny fluctuations that later made the formation of galaxies possible.

We shall never be able to see what the universe was like when just born.
In the first 380,000 years after Big Bang cosmos was completely opaque to light. There were lots of photons around that however will never be able to 'give light' to our instruments of observation: they were all the time absorbed by a sea of free-moving and frantic electrons in the very hot primeval universe.

The mass of the universe is mostly due to a type of matter still unknown.
Astronomers are convinced that the matter that is familiar to us is only a small part of the mass of the universe; the rest is 'dark matter' that does not emit radiations and therefore is not directly observable. Only the gravitational effects due to this invisible matter can justify the form of galaxies observed and their deployment in clusters.

G.F. Bignami, *We are the Martians*,
DOI 10.1007/978-88-470-2466-3_11, © Springer-Verlag Italia 2012

Stars are factories in which chemical elements are manufactured.

Every star, when it is born, is principally made of hydrogen and helium, the 'nuclear fuel' that will have to last her all her life. The star radiates energy because with nuclear fusion it 'burns' hydrogen and hellion, turning them into more massive atomic nuclei. Thus, the heavier chemical elements have originated.

Chemical elements heavier than iron are mainly formed when stars explode.

The process of nuclear fusion cannot generate nuclei more massive than iron. Chemical elements heavier than iron (for instance, copper, mercury, gold and uranium) are mainly formed when the bigger stars, in dying, implode, thus producing the huge stellar explosion called *supernova*.

We, like everything that surrounds us, are made of stardust.

Mendeleev's table, which orders chemical elements according to their periodical properties, is a huge achievement of the human mind. Just as wonderful it is that we understood that almost all the elements in the table—except hydrogen and helium—have been synthetized from stars.

One part of us dates straight back to Big Bang.

More than half of our body is made of water and every molecule of H_2O contains two atoms of hydrogen, the lightest chemical element in Mendeleev's periodic table. All the hydrogen which is present today in the universe was formed in the first 3 min after Big Bang.

It is not by chance that the Solar System was formed billions of years after the Big Bang.

The first stars that were formed after the birth of the universe were practically made up only by the matter that was created during Big Bang (80% hydrogen and 20% helium).

Further generations of stars—during the 8 billion years that passed before the formation of the Sun—have gradually enriched the universe with heavier chemical elements.

And a planetary system like ours can be formed only starting from a 'metallic' interstellar cloud that is rich in heavy elements.

Stars and their planets are formed practically at the same time.
When a metallic interstellar cloud gives rise to a star, the latter is surrounded by a disc of dust, residual from the cloud, out of which planets are formed in a few million years' time, a very short interval of time on the scale of cosmic events.

Astronomers today do not study celestial bodies only at a distance, but also by 'tasting' them directly.
Contact astronomy is based on the analysis of meteorites, bits of the Solar System that come to see us at home, and on the instruments of the probes that we send to touch other celestial bodies (planets, satellites, asteroids, and comets), sometimes even bringing a few bits back to Earth.

Every year various hundredweights of Martian rocks fall on Earth.
These meteorites, residual from ancient impacts of either asteroids or comets on the surface of Mars, arrive on our planet after wandering in space, sometimes for millions of years.

In the same way many bits from the Earth—residual from ancient collisions undergone by our planet—surely go coasting around the Solar System and some of them end up by falling on Mars. A similar 'interplanetary exchange' also takes place with the Moon and with Venus.

Comets tell the story of the origins of the Solar System.
Billions of comets that revolve at a huge distance from the Sun have not undergone the effects of the warming and of the solar wind, and thus they are still untouched witnesses of the time in which our planetary system was formed. Studying them in situ is the aim of the European mission Rosetta.

Recently humans have 'polluted' the Solar System with terrestrial microorganisms.
Our exploring probes, especially at the beginning of space research, did not get accurately sterilized before launch.

Thus, hosts of bacteria have been sent into space and on several celestial bodies that the probes touched upon.

Mars could harbour today, or had harboured in the past, forms of life.

The exploration of the red planet has shown that in the past on its surface liquid water ran and even today, in the slight atmosphere of Mars, methane is present, which could originate from the metabolism of terrestrial bacteria.

Even in the frozen outermost regions of the Solar System, there are environments potentially favourable to life.

Titan, Saturn's biggest satellite, is the only moon in the Solar System to have a thick atmosphere which is rich in organic molecules. A much smaller Saturn's moon, Encelado, may harbour water in the liquid state. And an underground moving ocean probably exists on Europa, a big moon of Jupiter.

Life is curiously asymmetrical on the molecular scale.

Many organic molecules exist in two forms, one the mirror image of the other, called levogyre and dextrogyre. In theory the two forms should be equally abundant, but, *on* the contrary—nobody knows why—it is not so in terrestrial organisms: in amino acids the levogyre form prevails, and in sugars the dextrogyre.

The same mysterious asymmetry has been observed in amino acids coming from outer space on board meteorites.

In the Solar System amino acids are found even very far from Earth.

In 2004 the probe Stardust visited comet Wild 2, gathered a little dust from its coma and brought it back to Earth. In the comet dust, glycine was found, the simplest among the amino acids employed by terrestrial living organisms to assemble proteins.

It is not sure that the first 'bricks of life' were formed on Earth.

On our planet every year thousands of tons of extraterrestrial materials fall. Many meteorites have revealed themselves rich in organic molecules, amino acids in particular, with chemico-physical characteristic identical to those of terrestrial living organisms.

It is thus possible that in the remote past the 'bricks of life' arrived on Earth from space: maybe we are the Martians.

We are looking for alien life also by means of 'galactic interceptions'.

For the past 50 years Project SETI (*Search for ExtraTerrestrial Intelligence*) has been active, which employs radiotelescopes to try and identify—up to the present with no success—coded messages sent to us by extraterrestrial civilizations.

Everybody can contribute to analyse the data taking part with their own computer by the diffuse calculation system SETI@home.

References

Atkins, P.W.: The Periodic Kingdom: A Journey into the Land of the Chemical Elements. Basic Books, New York (1995)

Bignami, G.F.: L'esplorazione dello spazio, Il Mulino (2006)

Bignami, G.F.: La storia nello spazio, Mursia (2001)

Crick FHC, Orgel LE (1973) Directed Panspermia. Icarus 19:341–346, http://profiles.nlm.nih.gov/SC/B/C/C/P/_/scbccp.pdf

Davies, P.: The Goldilocks Enigma: Why Is the Universe Just Right for Life. Penguin (2007)

De Bernardis, P.: Osservare l'universo, Il Mulino (2010)

De Duve, C.: Vital Dust: The Origin and Evolution of Life. Basic Books, New York (1995)

De Grasse Tyson, N., Goldsmith, D.: Origins: Fourteen Billion Years of Cosmic Evolution. W.W. Norton & Company, New York (2005)

Dyson, F.: From Eros to Gaia, Pantheon, New York (1992)

Flammarion, C.: Les terres du ciel, Didier et Cie (1877)

Flammarion, C.: Popular Astronomy: a General Description of the Heavens. Chatto & Windus, London (1894)

Genta, G.: Incontri lontani. Lindau, Torino (2009)

Guaita, C.: I pianeti e la vita. Gruppo B (2009)

Hoyle, F., Wickramasinghe, N.C.: Lifecloud: The Origin of Life in the Universe. J.M. Dent and Sons, London (1978)

Hoyle, F., Wickramasinghe, N.C.: Diseases from Space. Harper & Row, New York (1980)

Hoyle, F., Wickramasinghe, N.C.: Evolution from Space. Simon & Schuster, New York (1981)

Kamenetskii, M.D.F.: Unraveling DNA: The Most Important Molecule of Life. Basic Books, New York (1997)

Luisi, P.L.: The Emergence of Life. Oxford University Press, Oxford (2006)

G.F. Bignami, *We are the Martians*,
DOI 10.1007/978-88-470-2466-3, © Springer-Verlag Italia 2012

Maddox, J.: What Remains to Be Discovered: Mapping the Secrets of the Universe, the Origins of Life, and the Future of the Human Race. Free Press, New York (1998)

Regis, E.: What is Life? Investigating the Nature of Life in the Age of Synthetic Biology. Oxford University Press, Oxford (2008)

Schiaparelli, G.V.: La vita sul pianeta Marte. Associazione Culturale Mimesis (1998)

Schrödinger, E.: What Is Life? The Physical Aspect of the Living Cell. Cambridge University Press, Cambridge (1944). http://whatislife.stanford.edu/LoCo_files/What-is-Life.pdf

Villani, G.: Complesso e organizzato. Franco Angeli (2008)

Ward, P.D.: Life as We Do Not Know It: The NASA Search for (and Synthetis of) Alien Life. Viking, New York (2005)

Weinberg, S.: The First Three Minutes: A Modern View of the Origin of the Universe. Basic Books, New York (1977)